Biosensors: Microelectrochemical Devices

Sensors Series

Series Editor: **B E Jones**

BIOSENSORS: MICROELECTROCHEMICAL DEVICES

M Lambrechts and W Sansen

Katholieke Universiteit Leuven, Belgium

CRC Press
Taylor & Francis Group
Boca Raton London New York

CRC Press is an imprint of the
Taylor & Francis Group, an **informa** business
A TAYLOR & FRANCIS BOOK

CRC Press
Taylor & Francis Group
6000 Broken Sound Parkway NW, Suite 300
Boca Raton, FL 33487-2742

First issued in paperback 2019

ISBN-13: 978-0-367-40288-4

Library of Congress Cataloging-in-Publication Data

Catalog record is available from the Library of Congress

Visit the Taylor & Francis Web site at
http://www.taylorandfrancis.com

and the CRC Press Web site at
http://www.crcpress.com

To Annemie and Hadewych

CONTENTS

PREFACE

Sensors are becoming more and more important as they provide the interface between the real world and the world of process control, digital computers and data acquisition systems. Physical sensors for measuring pressure and temperature are already extensively established in industrial as well as in consumer applications. The use of chemical sensors on the other hand is rather limited. Only pH electrodes and conductivity sensors are well accepted in the daily practice of process control.

Biosensors, or sensors that are based on the use of biological material for their sensing function, were until a few years ago only used in research or clinical laboratories. A major breakthrough in the commercial world of biosensors came with the introduction of the 'ExacTech' system: a disposable glucose sensor produced with screen-printing techniques. This new trend, the realization of biosensors with microelectronic production techniques such as screen printing or photolithography, is the only way to realize biosensors in a cost-effective way. Several companies have recognized this economic fact and have also started their development programs. The same phenomenon has been seen previously with the introduction of mass-produced pressure sensors based on micromachining techniques in silicon.

Electrochemical biosensors realized with microelectronic production techniques (or *microelectrochemical* sensors) are the main topic of this book. The intention of the authors is the publication of an interdisciplinary book that can be read with a minimum of background knowledge by electronic engineers as well as by chemists or other interested researchers. Therefore, general principles such as basic electroanalytical chemistry and standard microelectronic procedures are briefly reviewed in this work. The content of this book is based on the doctoral thesis of Marc Lambrechts, entitled 'Planar voltammetric sensors'.

It is not the intention of the authors to provide a book which provides full details of all aspects of electrochemical sensors. This type of completeness is simply impossible. The basic idea of this book is to describe the necessary theory and technology for the realization of biosensors with microelectronic production techniques as seen by electronic engineers.

This book is thus intended for all persons who are intrigued by the interdisciplinary world of biosensor production technology. This includes engineers as well as chemists, material scientist or biotechnology researchers at senior or graduate level. It is especially written for these researchers who are working on the biological part of biosensors and want to know how to convert their sensor principle into a more commercial device by the use of microelectronic production techniques.

Marc Lambrechts and Willy Sansen
1 March 1992

ACKNOWLEDGEMENTS

We wish to express our gratitude to the many individuals who have contributed their ideas, their time and their energy toward the creation of this book. In particular, we wish to thank the following:

Jan Suls, Annick Claes, Guido Huyberechts and Paul Jacobs—colleagues in the ESAT–MICAS chemical sensor research group at the Katholieke Universiteit Leuven. Jan has taught us nice things about chemistry. Annick has spent a lot of time with glucose sensors. Guido was so kind critically reviewing this text and the Paul's enthusiasm was, and still is, contagious for the entire team.

Tin Kuypers (Dråger Medical Electronics, Best, The Netherlands) gave us a better insight into the world of biomedical sensors. Many of the ideas born during our discussions are reflected in this book.

Professor Nico F de Rooij (IMT, Neuchatel) and Dr E P Honig (Philips NatLab) provided us with material incorporated in this book.

Professor J M Kaufmann (Université Libre de Bruxelles), Dr John Pritchard, Johan Notré, Paul Jacobs, Eric Peeters and Jan Suls were so kind to proof read parts of the final text of this book.

CHAPTER 1

AN INTRODUCTION TO MICROELECTROCHEMICAL SENSORS

Sensors have become so important and so interesting because some of them allow us to extend and complete our own sensing capabilities. Moreover, sensors have become vital for control systems and robotics in general. Indeed, sensors are as important for modern electronic control systems as our own five main sense organs are for ourselves. Without our hearing, touch, smell, taste and vision sensing capacities we are vulnerable and limited in our capabilities. Sensors can best be seen as electronic equivalents for our sense organs. It is well understood however that the development of sensors is a quantum leap behind the realization of microelectronic devices. This is especially true for the sense organs that detect chemicals. On the other hand, nowadays high-resolution CCD cameras can provide vision for computers and extremely sensitive microphones can provide auditory information. To replace the touch sense, very precise temperature, force and pressure sensors exist; finally, tactile sensing is still in its infancy.

Electronic equivalents for the chemical senses, namely the taste and smell sense, do not exist at this moment. Also in our own biological systems, such as the immunological system and the glucose metabolism, chemical concentrations are measured and regulated with high precision. In order to replace or monitor these biological systems, extremely sensitive and selective sensors are necessary. Commercial chemical sensors are bulky, need regular calibration and are not selective and sensitive enough. However, if biological material is combined with existing chemical sensors into a biosensor, a similar selectivity and sensitivity as in biological systems can be obtained. Extensive research on this topic started a decade ago and has now resulted in several commercial sensors.

The best known example of a biosensor is without doubt the glucose sensor. As there is an extensive need for cheap disposable glucose sensors, the development of microelectrochemical glucose sensors with microelectronic techniques is the main topic in this book. Needle-type sensors are

1

not discussed: they are too difficult to make, their characteristics are not sufficiently reproducible and hence they are expensive. The technology of microelectronics, on the other hand, has shown to be able to provide highly reproducible devices at low cost. Mass-fabrication techniques have been in development for over two decades now, so sensor elements are discussed which mainly use these planar technologies.

In the following sections of this introduction, the basic principles and terminology related to this work are detailed. In section 1.6, the structure of this book is described.

1.1 SENSORS, BIOSENSORS, BIOPROBES, TRANSDUCERS AND ACTUATORS; A DISCUSSION OF BUZZWORDS

Classification of sensors and related devices is a very popular activity, as usually at least one paper at each sensor conference is devoted to this topic. Classifications have been made according to signal domains (Middelhoek and Noorlag 1981, Middelhoek and Audet 1987), electronic components (Bergveld 1986), transduction modes (Janata and Bezegh 1988), applications (Kobayashi 1985) and biological criteria (Aizawa 1983, Lewis 1985). Because every classification has its exceptions and problems, there is some confusion over terminology. The following clarification of sensor jargon gives an overall view and is of course also subject to discussion.

1.1.1 Transducers, sensors and actuators

According to etymology, the word *transducer* is derived from the Latin verb 'transducere–traducere' which means 'to transfer–to translate'. Therefore, a device that transfers or translates energy from one kind of system to another, in the same or another form, is termed a transducer. In the instrumentation environment, a transducer is used to indicate a device that transfers a signal from one energy form into a signal in another energy form. The energy forms can be electrical, mechanical, optical, thermal, magnetic or radiant (Middelhoek and Noorlag 1981). A device where the input and the output belong to the same signal domain can be identified as a *modifier* (e.g. a transistor).

A *sensor* can now be defined as a transducer that converts a signal of some specific form into an electrical signal. An *actuator*, on the other hand, can be defined as a transducer that converts an electrical signal into a signal of another form, usually a mechanical signal (see also figure 1.1).

Examples of sensors are pressure sensors, pH sensors and phototransistors. Examples of actuators are solenoids, piezoelectric devices and laser

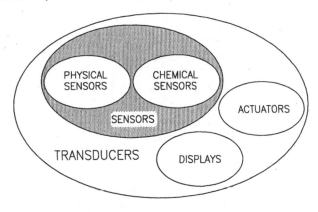

Figure 1.1   Classification of transducers according to the application.

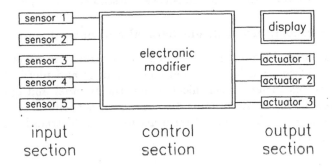

Figure 1.2   Schematic of a universal control system consisting of sensors, an electronic modifier, actuators and a display.

diodes. An electrode at which hydrogen or oxygen is generated by applying a potential is an example of a chemical actuator. A display is a special kind of transducer that converts an electrical signal into a readable form; a CRT display, LCD screen or LED array are typical display devices that can be found almost everywhere nowadays.

A universal electronic control system can be represented as consisting of sensors to convert input parameters into an electrical signal; these electrical signals are then processed by an electronic modifier, e.g. a microprocessor (see figure 1.2). The output of the modifier is converted by actuators into suitable stimuli or shown on a display. In this work attention is focused on the input section, i.e. the sensors. Miniature versions of the universal control system shown in figure 1.2 can be used as an implantable replacement for vital functions. Such an Internal Human Conditioning System (IHCS) suited for 'in vivo' applications is for example described in Sansen (1982).

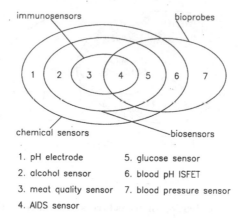

1. pH electrode   5. glucose sensor
2. alcohol sensor   6. blood pH ISFET
3. meat quality sensor 7. blood pressure sensor
4. AIDS sensor

Figure 1.3   The difference between biosensors and bioprobes.

1.1.2   Chemical, physical, mechanical and optical sensors

Sensors can be divided further according to the signal domain of the input signal. A division can be made between chemical and physical sensors. Physical sensors include mechanical, magnetic, thermal and optical sensors. In this book, all attention is directed towards chemical sensors. The chemical sensors can be divided further according to the transduction mode (see section 1.2).

1.1.3   Biosensors and bioprobes

A special type of chemical sensor is the *biosensor*. A biosensor is defined as a sensor that makes use of biological or living material for its sensing function. A glucose sensor is the best known example of a biosensor. The detection of glucose is based on the enzyme glucose oxidase, a biological component usually extracted from *Aspergillus Niger* (another source of this enzyme can be *Penicillium notatum*). An *immunosensor* is a biosensor that is based on immunological components. A *bioprobe* is defined as a sensor that measures vital functions of living beings. A blood pressure sensor is a typical example of a bioprobe. More examples of biosensors and bioprobes can be found in figure 1.3.

Biosensors are a subdivision of chemical sensors. Biosensors that measure immunological parameters are called immunosensors. It is important to notice that a sensor can be a biosensor as well as a bioprobe. An AIDS sensor is for example an immunosensor that is based on biological material, so it is a biosensor. Since it monitors the immunological system—a vital function—it is also a bioprobe. The same can be said about glucose sensors; however, if an enzymatic glucose sensor is used for monitoring an enzymatic reaction in a fermentation tank, then the glucose sensor is still a biosensor but not a bioprobe. A good example of an immunosensor that is not a bioprobe is a meat quality sensor. A well known example of a chemical sensor that is not a biosensor is the classical pH glass electrode.

1.1.4   Solid-state, integrated, planar, smart and intelligent sensors

The term *solid-state sensor* is used to indicate that the operation of the device depends on effects and phenomena situated in a solid. Piezoresistive pressure sensors and solid/electrolyte gas sensors are typical examples of solid-state sensors. Sometimes, the term solid-state sensor is used incorrectly, indicating a silicon-based sensor.

Over the last few years, great interest in mass-produced sensors has arisen because of the low production cost (see also section 1.3). Therefore, semiconductor or microelectronic technology is the pre-eminent fabrication technique, as this technique is well known and allows extreme miniaturization.

A sensor is an *integrated sensor* if the sensing operation is based on a direct influence on an electronic component in the silicon or other semiconductor material; the sensing function is integrated with the microelectronic component. The electronic component can be a resistor, a MOSFET, a bipolar transistor or a capacitor. Integrated sensors are nowadays commercially available as, for example, piezoresistive pressure sensors. The ISFET is also a good example of an integrated sensor.

If the sensor is made using microelectronic production techniques, but does not require the intrinsic characteristics of the semiconductor material, then this sensor is termed a *planar sensor*. The silicon wafer is in this case only a flat substrate and can in principle be substituted by a quartz or a glass substrate. Sensors made with thick-film technology are also planar sensors. Practically all chemical sensors made with microelectronic techniques are planar sensors (ISFETs excepted). In general, it can be stated that integrated sensors are made *in* silicon whereas planar sensors are made *on* silicon.

The combination of interface electronics and an integrated or planar sensor on one chip results in a so-called *smart sensor*. This requires that at least some basic signal conditioning is carried out on the sensor chip. A synonym for smart sensor is *intelligent sensor*. The major advantage of smart sensors is the improved signal-to-noise and electromagnetic interference characteristics of the device. Also, temperature compensation and calibration routines can be build into the sensing device itself.

Now that the terminology is clarified, the subject of this book can be defined as planar biosensors and bioprobes. The sensing principle, namely electroanalytical chemistry, is detailed further in the next section; the production technique and the applications are introduced in section 1.3 and 1.4, respectively.

1.2   ELECTROCHEMISTRY: A POWERFUL ANALYTICAL TOOL

The point of interest of this book is the realization of cheap planar chemical sensors. From the economic point of view, the most interesting sensors are gas sensors and biomedical sensors. These sensors have a large market, as they are applicable to consumer applications. A reliable gas sensor for CO, natural or town-gas can be installed in houses and prevent CO intoxication or explosions. Taking into account the large amount of houses heated by combustion processes, a large amount of sensors are required to protect these houses. However, a long sensor lifetime is required. With the current status of technology, cheap gas sensors with a long lifetime and a good selectivity can not yet be produced. However, recent results in gas sensor research indicate that cheap devices will be available within a few years.

Biomedical sensors are, from the economic point of view, even more interesting. Not only are a large amount of sensors needed, but they should preferably be disposable to prevent re-sterilization problems and to avoid the risk of transfer of diseases from one patient to another. As they are intended for single-use applications, only a short operational lifetime of the sensor is required. The price at which medical sensors can be marketed is also higher than for industrial or consumer devices. In the medical world, the most important sensors are glucose sensors and sensors for the determination of dissolved oxygen in blood (see also section 1.4).

For chemical sensors, different transducing principles are available: gravimetric, thermal, optical and electrochemical principles can be used to measure chemical concentrations (see also table 1.1).

Gravimetric sensors are based on the measurement of a change in mass. As detector elements, surface acoustic wave devices (SAW) or piezoelectric crystals are used. These detector elements provide a very high resolution. Changes in mass of the order of magnitude of one picogram can be detected. The operation of these oscillating devices in a liquid environment is, however, questionable.

Temperature is, of course, the physical quantity measured with a thermal sensor. As detector elements, thermistors or pyroelectric devices are used. Thermistors are cheap devices which are commercially available in a great variety of sensitivities and packaging. The principle of operation is based on the measurement of the temperature difference due to an exothermic or endothermic enzymatic reaction.

Optical sensors make use of the effect of chemical reactions on optical phenomena. Photodiodes, phototransistors and interferometers in combi-

Table 1.1 Transducing principles for chemical sensors (gravimetric, thermal, optical and electrochemical).

signal domain	gravimetric	thermal	optical	electrochemical
measured quantity	mass	temperature	light intensity, reflectance, interference, fluoresence	potential, current, conductance
detector	SAW, piezoelectric crystals	thermistor, pyroelectric devices	photodiode, SPR, interferometer	noble metal electrodes, ISFETs
examples and applications	immunosensors, gas sensors	enzyme sensors	oxygen sensors enzyme sensors affinity sensors	ISEs, glucose sensor, Clark cell, pH electrode

nation with optical fibres are typical detector structures. A nice example of an optical biosensor which has recently been introduced on the market is the BIAcore system from Pharmacia Biosensor AB (Pharmacia 1991). This system is based on surface plasmon resonance (SPR) techniques for real-time biospecific interaction analysis. Primarily, it is intended as a research tool; the sensing unit is a re-usable sensor chip. Other research groups are working on single-use SPR systems (Attridge *et al* 1991). Another interesting examples of optical biosensors are the capillary-fill devices (Deacon *et al* 1991). Optical techniques seem to be of interest especially for affinity-based immunosensors.

Electrochemical sensors make use of electrochemical reactions. As a parameter, the potential or conductance between two electrodes or the current through a polarised electrode is measured. A great variety of chemical substances can be measured in this way.

When taking a close look at papers on chemical sensors published from 1985 to 1987 (Janata and Bezegh 1988) and when comparing with the number of papers in the period 1988 to 1989 (Janata 1990), it is seen that the majority of the papers deal with electrochemical sensors, namely conductometric, potentiometric and voltammetric (or amperometric) sensors. As indicated in figure 1.4, the interest in thermal sensors is constant, optical sensors and gravimetric sensors are rising and electrochemical sensors seem to enjoy constant interest. Ion-selective electrodes show a major decrease in interest, whereas optical sensors are the fastest growing area in scientific literature on sensors.

It is important to notice that voltammetric devices are the most important electrochemical sensors (not taking into account the potentiometric ion-selective electrodes (ISEs), as they form a special kind of device). It is also interesting to see that a large amount of papers published on chemical

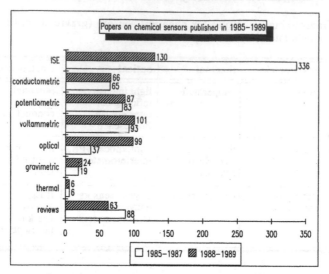

Figure 1.4   Trends in the field of chemical sensors according to the amount of papers published on the different topics (based on Janata (1990)).

sensors are review papers.

As will be discussed in chapter 2, the important principles of electroanalytical chemistry are the following.

1. Potentiometry

Potentiometric sensors are based on the measurement of the *potential* at an electrode in a solution. This potential is measured in an equilibrium situation, i.e. no current is allowed to flow during the measurement. According to the Nernst equation, the potential is proportional to the logarithm of the concentration of the electroactive species.

2. Voltammetry

Voltammetric sensors are based on the measurement of the *current-voltage relationship* in an electrochemical cell consisting of electrodes in a solution. A potential is applied to the sensor and a current proportional to the concentration of the electroactive species of interest is measured. Amperometry is a special case of voltammetry where the potential is kept constant as a function of time.

3. Conductometry

Conductometric sensors are based on the measurement of the *conductance* between two electrodes in a solution. The conductance is measured by applying an AC potential with a small amplitude to the electrodes in order to prevent polarization. The presence of ionic elements are detected as an increase in the conductance.

In this book, well known electroanalytical techniques are described for sensors fabricated with microelectronic production techniques. This combination of microelectronics and electrochemistry results in a new discipline that can be designated 'microelectrochemistry'. The resulting devices—'microelectrochemical devices'—have exciting characteristics, not only in the field of electroanalytical chemistry or the medical world, but also from the economic point of view. As a result, the content of this book is expected to remain relevant for quite some time.

1.3 'PLANAR' SENSORS: AN ECONOMIC FACT

Chemical sensors realized with microelectronic production techniques have several advantages. The majority of these advantages have an economic foundation. For example, sensors suitable for the consumer market can best be produced with microelectronic-based mass-production techniques, resulting in cheap planar sensors. The major advantages of planar sensors are as follows.

1. Low cost per sensor
As microelectronic production techniques are essentially mass-production techniques by definition, they allow production of low-cost sensors. However, this requires that all process steps necessary for the realization of the complete planar sensors are based on mass-production techniques, packaging included. Every single manual process step will result in an extensive increase of the cost per sensor. In order to reduce the processing and development cost, it is preferable to make use of as many standard process steps as possible. Standard processing steps, such as those required for CMOS or bipolar processes, are well defined and relatively inexpensive. Every non-standard process step will also increase the cost per sensor.

However, it is important to notice that a low cost can only be reached when a huge amount of similar sensors can be marketed. If the market only demands a small quantity of identical sensors, microelectronic techniques result in a high cost per sensor due to high development costs and the large investment required. An interesting alternative is then the realization of sensors with thick-film technology.

2. Small dimensions
With microelectronic techniques, very precise structures can be realized. Nowadays features with a line-width smaller than 1 μm are common place in VLSI integrated circuit factories, so elements with a very high precision can be realized. As the physical phenomena occurring at these small elements are quite often different than those for their larger counterpart, devices with novel characteristics can be made.

Small dimensions are also an essential requirement for 'in vivo' biomedical sensors, as they have to be implanted into the body. Although it is possible to produce implantable needle-type sensors, only planar sensors can be produced in large quantities at a low cost.

3. High reproducibility

Planar sensors are always produced in batch processes. This batch processing in combination with the high precision of microelectronic techniques results in an improved reproducibility of the sensor. The ultimate goal is the development of a sensor with such a degree of reproducibility that calibration is not necessary. Calibration of sensors is carried out on the individual sensors and is, moreover, expensive. It is clear that the cost of the sensor will be reduced if calibration can be avoided.

4. Possibility of smart sensor realizations

If the processing of the sensor is made compatible with standard IC techniques, such as CMOS or bipolar processes, then interface electronics can be integrated with the sensor on-chip. From the economic point of view, it is preferable that the sensor processing is carried out after the completion of the standard circuit processing. In this way, the interface circuits can be processed in a silicon foundry and the sensor manufacturer does not have to bother with the development of a complete IC process.

With smart sensors a better signal-to-noise ratio can be realized and the electromagnetic interference can be reduced. The characteristics of the sensor can be improved by the elimination of non-ideal behaviour, such as non-linearity, cross-sensitivity, temperature sensitivity, offset and drift. With interface electronics, on-chip signal conditioning can also be performed. Evident examples are analogue-to-digital conversion, current-to-voltage conversion, signal multiplexing and impedance transformation. It will ultimately be necessary to provide smart sensors with a universal interface bus structure (Henning 1991). In this way, a system developer can use the same interface structure to allow the communication of different types of sensors with a digital control unit. Easy integration of sensors in systems has then become reality.

In the world of mechanical sensors, smart devices are already well established. Smart mechanical sensors are common commercial devices and gate arrays specially designed for the realization of smart sensors are under development (Haviland *et al* 1991).

1.4   ON THE NEED FOR GLUCOSE AND OXYGEN SENSORS

In the biomedical world, there is an extensive need for reliable sensors for a great variety of chemicals. The most important sensors for the control of vital functions are without doubt glucose sensors for diabetic patients and dissolved oxygen sensors for use in intensive care and during anaesthesia.

Diabetes mellitus is a chronic disease in which the blood glucose concentration is too high. In normal healthy patients the glucose concentration is regulated within a narrow range (3.5–6.5 mM) by the pancreas. This is achieved by insulin secretion from the β-cells of the pancreatic islets of Langerhans. Insulin is a protein hormone that lowers the glucose concentration in blood by suppressing glucose output from the liver and by stimulating glucose uptake into muscles and tissue.

In Europe and North America, 1–2% of the population suffers from diabetes. World-wide, it is estimated that 30 million patients are diabetic (Pickup and Rothwell 1984). Two types of diabetic patients can be distinguished.

1. Type 1, insulin-dependent or juvenile-onset diabetes

Type 1 diabetic patients comprise about 20% of the diabetic population. In Belgium approximately 35 000 patients (of 10 million inhabitants) suffer from this disease. The onset of type 1 diabetes is situated in childhood and early adulthood. These patients suffer from β-cell injury, resulting in low or zero blood insulin concentrations and therefore high blood glucose concentrations. The exact cause of the β-cell injury is not known; currently, it is believed that the tendency to develop diabetes is inherited and that additional environmental triggers, such as viruses or toxic chemicals, are necessary in later life for the development of the actual disease (Turner 1985). These patients must receive daily insulin injections to remain alive.

2. Type 2, non-insulin-dependent or maturity-onset diabetes

About 80% of the diabetic population have type 2 diabetes. The onset of the disease is situated between 50 and 70 years of age: the type 2 patients are usually obese. The β-cells of these patients are still intact and blood insulin concentrations are normal or elevated. High blood glucose concentrations in type 2 diabetic patients are due to a failure of insulin to act properly at the target tissues or are due to a failure of the β-cells to secrete sufficient insulin for the patient's needs. These patients are non-insulin dependent; usually, blood glucose concentration can be regulated to acceptable levels by a proper diet and by oral hypoglycaemia agents.

The blood glucose level in a diabetic patient varies considerably. Values between 1 to 30 mM may be observed. This range is defined as the clinically significant range for glucose sensing. A high blood glucose concentration is named hyperglycaemia; a low glucose concentration is indicated as hypoglycaemia. Both abnormal conditions result in severe short- and long-term complications.

Hypoglycaemia is the greatest fear of type 1 diabetic patients, especially during sleep. Hypoglycaemia results in mental confusion, as glucose is the main energy source of the brain. Prolonged hypoglycaemia leads to coma, brain damage and death.

Hyperglycaemia manifests itself by thirst and frequent urination; if untreated by insulin injections, the patient becomes severely ill as the blood glucose and ketone body concentrations rise. The patient dehydrates, becomes acidotic and sleepy and eventually goes into a coma or dies. This syndrome is known as diabetic ketoacidosis. In the long term (> 10 years), hyperglycaemia complications appear. Small blood vessel disease results in kidney failure and blindness; large blood vessel disease causes accelerated coronary artery problems, strokes and circulatory impairment in arms and legs. Also nerve damage can occur.

In current treatments of diabetic patients, insulin injections and diets are regulated with self-monitoring techniques according to meals, exercise and daily blood glucose analysis. For some years, implantable insulin delivery systems have existed (Spencer 1981), but, as these systems are open-loop devices, an accurate control of blood glucose levels can not always be achieved.

However, it is believed that with improved blood glucose control in diabetic patients, the long-term complications of this chronic disease can be eliminated. A reliable blood glucose sensor will thereby substantially improve the quality of life of diabetic patients. Six different types of applications or devices can be defined for this purpose.

1. An 'in vitro' sensor for self-monitoring can replace the paper strips impregnated with chemicals that result in a colour variation according to the glucose concentration.

2. An 'in vitro' bench-top blood glucose analyser with a cheap disposable glucose sensor can for example be used in the office of the physician for precise glucose determinations.

3. An 'in vivo' hypoglycaemia alarm that warns the patient of low blood glucose concentrations can eliminate the risk of coma or death during a nocturnal or unnoticed hypoglycaemia.

4. An 'in vivo' hyperglycaemia alarm can give an early warning of rising glucose concentrations so that the patient can take the necessary precautions to avoid ketoacidosis.

5. An 'in vivo' glucose sensor for short-term bedside monitoring can help physicians in determining the insulin dependence of new diabetic patients and is a valuable tool for research on the glucose metabolism.

6. An 'in vivo' glucose sensor for a closed-loop insulin delivery system will allow the replacement of the pancreas with an artificial organ and result in an improved and stable glucose level.

Basically, three types of glucose sensors are necessary to fulfil the needs described above:

1. a disposable, single-use 'in vitro' glucose sensor,
2. an implantable 'in vivo' glucose sensor for short-term use,
3. an implantable 'in vivo' glucose sensor for long-term use.

Table 1.2 Specifications of three different types of glucose sensors according to the applications.

sensor type	"in vitro"	"in vivo"-short term	"in vivo"-long term
applications	home-monitoring lab analyser	bed side monitoring	artificial pancreas hypoglycaemia alarm hyperglycaemia alarm
appearance	dip-stick	catheter-tip, needle	catheter-tip
dimensions	5 mm x 10 mm	0.7 mm x 5 mm	0.7 mm x 5 mm
lifetime	< 1 day	1-3 days	> 1 year
linear range	1 - 30 mM	1 - 30 mM	1 - 15 mM
accuracy	5%	5%	5%
biocompatibility	low	good	excellent
reliabilty	good	good	excellent
typical price	1 $	50 $	2000 $

The necessary specifications for these three types of glucose sensors are summarized in table 1.2. In the literature, glucose sensors have been reviewed by several authors: for example, Albisser (1979), Santiago *et al* (1979), Peura and Mendelson (1984), Cardosi and Turner (1987) and Pickup *et al* (1988).

In this book the development of planar glucose sensors for 'in vivo', as well as 'in vitro', use are described. The 'in vivo' sensors are intended as disposable glucose sensors with an operational lifetime of two days. The restricted lifetime (< 3 months) of the sensors does not allow a long-term 'in vivo' use. This limited lifetime is not due to device problems but to loss of activity of the used enzyme and biocompatibility problems; ongrowth of tissue causes drift. A lot of work from the biological point of view still has to be undertaken prior to the accomplishment of the final goal for all researchers on glucose sensors: *'the realization of an 'in vivo' glucose sensor with an operational lifetime of more than one year'*.

Glucose is not the only relevant parameter for physicians. The oxygen and carbon dioxide partial pressure and the pH of blood are of critical importance for the functioning of the human metabolism; these parameters are known as blood gas values. The oxygen partial pressure (pO_2) is an especially useful parameter since this reflects directly the functioning of the heart and lungs. The oxygen concentration is so important that, if possible, the physician would like to measure this parameter continuously. This can be achieved with an 'in vivo' oxygen sensor. To have a complete picture of the blood gas values the partial pressure of carbon dioxide (pCO_2) and the blood pH are usually measured as well.

Up to now the blood gas analysis technique is based on individual blood samples. An arterial blood sample is taken from the patient and analysed immediately with a bench-top blood gas analyser. This procedure typically takes 15 minutes and is rather expensive. The total cost of one complete blood gas analysis on one sample can be estimated at 50$. In the USA alone at least 25 million blood samples a year are taken from patients in order to determine pO_2, pCO_2 and pH (Parker 1987). Although this technique has saved a lot of lives and has given physicians valuable information, the procedure is very expensive and extremely slow in comparison with the possible changes in the blood gas values.

Nowadays very sophisticated systems exists. A nice example is the GEM-6 PLUS system from Mallinckrodt. The heart of this system consists of a sensor card with several electrochemical sensors on it (pH, pO_2, pCO_2, K^+, Ca^{++}, Na^{++} and haematocrit). The sensor card and the corresponding calibration solutions are contained in a disposable cartridge which is suitable for 50 samples over a maximum period of 72 hours (Mallinckrodt 1990). This kind of system is a major breakthrough in the field of blood gas analysis but does not yet give the same information as a real 'in vivo' sensor.

In figure 1.5, typical time constants for clinical measurements are indicated for the most significant parameters. The measurement of the oxygen partial pressure is without doubt the most time-critical. Intermittent analyses of blood samples are not sufficient for good control of the patient under critical conditions.

The measurement of blood gas values is particularly important for preterm infants which are very vulnerable to respiratory problems. Low pO_2 levels may result in brain damage and in death. With high pO_2 levels related to administration of oxygen, on the other hand, blindness can occur in the baby.

During surgery, good control of blood gas values could save many lives related to anaesthesia problems. During open heart surgery, the patient is attached to a heart–lung machine that temporarily takes over these vital functions; with an 'in vivo' oxygen sensor, the condition of the patient can be monitored continuously and can control the heart–lung machine. An 'in vivo' oxygen sensor can also find use in intensive care units for the continuous monitoring of critically ill patients. Finally, implantable defibrillators nowadays work without a haemodynamic parameter in the decision process concerning the administration of defibrillation shocks to the patient. An implantable oxygen sensor could serve here as a second parameter in parallel with the electrocardiogram.

From these introductory remarks on the necessity of glucose and oxygen sensors for biomedical applications, it can be concluded that there is an ex-

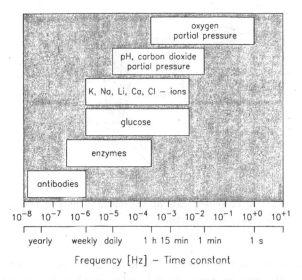

Figure 1.5   Typical time constants for clinical measurements during an intervention or treatment.

tensive and urgent need for these sensors. Taking into account that a large amount of small, disposable and cheap sensors is needed, it is clear that microelectronic techniques can provide a possible solution for the fabrication of these devices. In this book the necessary tools to integrate sensors with microelectronic devices are described.

1.5   GLUCOSE SENSORS AS A BASIS FOR MORE EVOLVED BIOSENSORS

A glucose sensor is the best known example of a biosensor. The realization of a glucose sensor is the beginning of the evolution towards more complex and more sensitive biosensors. The possibilities are outstanding.

In the introduction of this work, a biosensor is defined as a sensor that makes use of biological components for the sensing function of the device. As a general concept, a biosensor can be seen as consisting of two components: a receptor and a detector (figure 1.6). The receptor, i.e. the biological material, recognises selectively the material to be sensed and responds to this material by a change of an external physical parameter. This physical parameter is then monitored by the detector, resulting in a change of an electronic characteristic. In the well known example of the enzymatic glucose sensor (see section 2.3.2), the receptor is the enzyme glucose oxidase and the detector is a H_2O_2-sensitive voltammetric sensor. The enzyme converts glucose into H_2O_2 and this H_2O_2 is then converted into an electrical current. Therefore, the current in the detector is linearly

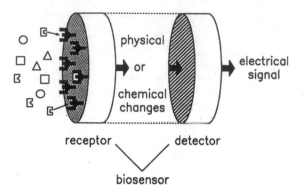

Figure 1.6   A biosensor can be represented as consisting of two parts: a receptor and a detector.

proportional to the glucose concentration. As discussed already for glucose sensors, a receptor can be used in combination with different detectors, and one detector can be used in combination with several receptors, resulting in a large variety of biosensors.

In table 1.3 an overview is given of the different detectors that can be used in the development of biosensors. It is important to notice that these detectors are not selective at all. Selectivity is obtained in the receptor; as biological materials such as enzymes can be extremely selective, very selective biosensors can be made. Not only enzymes are selective; other biological material can also be used as a receptor. An overview of the different receptors is shown in table 1.4.

The major problem with the measurement of clinically important parameters is the extremely low concentrations at which some important molecules are present in the body. Enzymes, drugs, hormones and immunological components such as antigens and antibodies are found in concentrations ranging from 10^{-8} M to 10^{-14} M and even lower. For the determination of antigens (e.g. a virus) the detection of a single unit of the antigen in the sample is the ultimate goal, as this single unit indicates that the patient has been infected by the virus. Single-unit detection is not possible with electrochemical techniques as this would involve for example the detection of a single electron in the case of a voltammetric sensor. The noise of electrochemical sensors, however, is too large to achieve this. Therefore chemical amplification has to be used to increase the concentration of an electroactive species. Several techniques for chemical amplification are known from biology as well as from clinical biochemistry. Interesting approaches are based on enzymes, cofactors, liposomes, lipid layers, DNA probes and bacteria. Hence, incorporation of chemical amplification has become a necessity in order to achieve the desired sensitivity and selectivity.

Table 1.3   An overview of possible detectors for biosensors.

detector	measurement principle	typical application
pH glass electrode	potentiometry	enzyme sensors
pH ISFET	potentiometry	enzyme sensors immunosensors
ion-selective electrode	potentiometry	enzyme sensors ion sensors
O_2-sensor	voltammetry amperometry	enzyme sensors microbial sensors
H_2O_2-sensor	voltammetry amperometry	enzyme sensors immunosensors
conductimeter	conductimetry	enzyme sensors
gas sensors	potentiometry voltammetry	enzyme sensors microbial sensors
fibre optic sensors	optical methods	ion sensors enzyme sensors immunosensors
thermistors	calorimetry	enzyme sensors microbial sensors
piezo-electric crystals	gravimetry	immunosensors
SAW- devices	gravimetry	immunosensors

For more details on these receptors, detectors and chemical amplification techniques, the reader is referred to the corresponding publications or to review papers such as Pace (1981), Aizawa (1983), Lowe (1985), Janata and Bezegh (1988), Turner *et al* (1987a).

1.6   AN INTRODUCTION TO THE STRUCTURE OF THIS BOOK

This interdisciplinary book describes the necessary elements that are required for the realization of planar biosensors or bioprobes based on electrochemical detection. Microelectronic production techniques are applied for the mass fabrication of the sensors. Basically, this book can therefore be dived into three parts: (1) introduction and electrochemical principles, (2) microelectronic fabrication technology and (3) case studies.

Chapter 1 gives an introduction on microelectrochemical sensors. After

Table 1.4   An overview of possible receptors for biosensors.

biological material	device name	measurement technique	typical analyte
enzymes	enzyme sensor	voltammetry potentiometry conductimetry calorimetry	glucose peniciline ureum
antibody-antigen	immunosensor	potentiometry voltammetry	hepatitis HCG
ionophores	ISE	potentiometry	K^+
bacteria	microbial sensor	voltammetry potentiometry	BOD nerve gases
tissue	tissue sensor	potentiometry	
lipid layers	bilayer membrane sensor	conductimetry potentiometry	

a discussion on sensor terminology, electroanalytical techniques are introduced as a well defined class of chemical sensors. The need for planar sensors is detailed and the major applications of microelectrochemical sensors are indicated.

In chapter 2, microelectrochemical sensors are placed in a historical context and the basic electrochemical principles, essential for the understanding of this book, are summarized. Here, electroanalytical chemistry is discussed from the point of view of an electronic engineer interested in electrochemical sensing principles applied to planar sensor technology. This chapter deals with electrode processes, potentiometry, voltammetry and conductometry. In order to allow comparison with microelectrochemical sensors, a special class of silicon-based chemical sensors is detailed, namely the ion-sensitive field effect transistors (ISFETs). Also in this chapter, the electrochemical reactions and cell configurations necessary for the realization of glucose and oxygen sensors are detailed. Because of the review-based character of this chapter, it can be skipped by researchers familiar with electroanalytical principles.

In chapter 3, the measurement techniques used for the evaluation of planar electrochemical sensors are reviewed.

Planar technology for sensor development is detailed in chapter 4. Because IC technology is an interesting base for a planar sensor process, standard IC technology and the related processing steps are outlined. This part of chapter 4 can be skipped by all microelectronic engineers familiar with CMOS processing. The second part deals with special microelectronic techniques applied to sensor applications. For the realization of planar voltammetric sensors, lift-off techniques, micromachining techniques in silicon and deposition procedures for planar chemical membranes are described. Thereafter, an alternative microelectronic production technique for planar voltammetric sensors is presented: namely, thick-film technology. Chapter four is concluded with a comparison of the different production techniques from an economic point of view.

In chapter 5, some case studies on microelectrochemical sensors are presented. Special attention is paid to a generic CMOS-compatible process for electrochemical sensors developed at the Katholieke Universiteit, Leuven (K U Leuven); different electrode configurations are detailed. Also, packaging techniques are revealed. The different electrode configurations and electrode materials are evaluated by means of the techniques listed in chapter 3. As the corrosion resistance of planar Ag electrodes is a critical problem, a special section is devoted to this topic.

In the second part of chapter 5, practical realizations of microelectrochemical sensors are detailed. Attention is paid to sensors developed by the authors, as well as by other research groups.

Thick-film voltammetric sensors are explained in chapter 6. With thick-film technology, interesting biosensors can also be implemented.

This book finishes with some concluding remarks and a list of references from the literature.

CHAPTER 2

BASIC ELECTROCHEMICAL PRINCIPLES

The earliest electrochemical experiments date from the first years of the nineteenth century when Volta (1800) demonstrated the interaction between electrical current in an electrolyte and chemical reactions. At that time he invented the Volta pile; the first battery. In 1834 Faraday formulated a quantitative relation between the total amount of current passed through an electrochemical cell and the amount of chemicals produced. This relation is known as Faraday's law. Faraday is also the originator of expressions such as electrode, cathode, anode, cation, anion,.... .

In 1879 Helmholtz proposed a simple model for the space-charge region surrounding the electrode; this model was based on a parallel-plate capacitor. In the beginning of the twentieth century, famous scientists explored a variety of electrochemical characteristics. Nernst derived his equation from the general thermodynamic equations proposed by Gibbs. Tafel (1905) found an experimental relation between the potential over the electrode interface and the current through that interface. The Nernst and Tafel equations are both a special case of a more general equation derived by Butler and Volmer. These laws and equations are discussed in section 2.1.1.

Goüy (1910) and Chapman (1913) improved the space charge region model of Helmholtz. This model was further refined by Stern in 1924. The current model of the space-charge region was introduced in 1947 by Graham and is known as the electrochemical double layer. These different models are detailed in section 2.1.2.

Practical applications soon emerged. Glass pH and reference electrodes came into general use ago, everywhere in the industrial and medical environment about 50 years. These electrodes are a direct implementation of the Nernst equation and are based on the measurement of the cell potential. For this reason this measuring technique is named 'potentiometry'. The theory and applications of potentiometry are detailed in section 2.2. In this section, special attention is paid to ion-selective electrodes (ISE) as these electrodes are of special interest for biomedical applications. Voltammetry is a more general technique which can be defined as the measurement of current–potential relationships of an electrode immersed in a solution containing electroactive species. The study of current–potential curves started at the end of the nineteenth century when Salomen (one of the co-workers of Nernst) in 1897 recorded current–potential curves for the electrolysis of silver ions.

The dropping-mercury electrode (DME) is well known for analytical applications. Its invention dates from the late 1920s and most of the text books on electrochemistry deal extensively with this technique. Although not accepted by all authors on electrochemistry, voltammetry at a dropping-mercury electrode is named polarography. Polarography is sometimes used incorrectly as a synonym for voltammetry; according to modern classifications, polarography is a subclass of voltammetry. Although the dropping-mercury electrode is of no direct interest for 'in vivo' biosensors, it is the basis of electroanalytical chemistry.

In 1956 Clark developed an amperometric sensor for the measurement of dissolved oxygen (Clark 1956). This device, known as the Clark cell, gave rise to a widespread use of electrochemical sensors for blood gas analysis. The Clark cell consists of a silver anode and a gold or platinum cathode covered with an oxygen permeable membrane. Clark cells, built into a catheter, are nowadays commercially available for 'in vivo' monitoring of dissolved oxygen in blood. These sensors are built using conventional wire and tube technology. However, it was soon discovered that a Clark cell can also be made with planar integrated circuit technology (Siu and Cobbold 1976) and several papers dealing with this topic have been published (Engels *et al* 1983, Sansen *et al* 1985, Sansen and Lambrechts 1985, Koudelka and Grisel 1985, Miyahara *et al* 1983, Suzuki *et al* 1988). The sensor described in Engels *et al* (1983) has now evolved into an 'in vivo' device in a pre-commercial phase, readied by Dräger Medical Electronics of Best in the Netherlands.

The Clark cell was also the source of inspiration for several other types of sensors based on the electrochemical reduction of oxygen. In combination with a glucose oxidase membrane, a glucose sensor can be made (Updike and Hicks 1967). For instance, the commercially successful Beckman glucose analyser is based on this principle.

This early work on glucose sensors has now evolved into a world-wide interest in biosensors based on electrochemical detectors as an interface between the chemical and the electrical world and on enzymes or other biological components as highly selective modifiers. The basics of voltammetry and the major biomedical applications are discussed in section 2.3.

Conductometry is not a very popular electrochemical technique, although it has some interesting possibilities. The main difference with conductometry and the other techniques is that conductometry deals with processes in the bulk of the solution whereas potentiometry and voltammetry deal mainly with processes at the surface or the near vicinity of the electrode. Theory and applications of conductometry are described in section 2.4

A completely different electrochemical sensor is the ion-sensitive field effect transistor, or ISFET, which was discovered by Bergveld in 1970 (Bergveld 1970). The idea of the ISFET is very attractive, as it can easily be miniaturized and combined with electronic circuits on the same silicon chip; an ISFET in its most simple form is in fact a normal MOS transistor without a gate. The electrolyte is in contact with the gate oxide, so the electrical parameters are modulated by variation of the chemical characteristics of the electrolyte (mainly by pH).

All over the world the ISFET has been investigated thoroughly and several laboratories in universities and industry have spent a lot of work and money on the development of reliable ISFETs. A large series of articles on the ISFET has been published, especially on the theoretical background of the device, since at the beginning the operation mechanism was not well understood. Despite all these efforts, no commercial products have appeared on the market yet, due mostly to drift problems in the device. The ISFET and its relation to microelectrochemical sensors is described in section 2.5.

The aim of this chapter is to give a short introduction to electrochemical principles so that the reader can follow the discussion about planar electrochemical sensors. It is not a replacement or enhancement of the standard textbooks on electrochemistry, rather an anthology of these books, as seen by a microelectronic engineer. The following books have been consulted: Adams (1969), Ewing (1975), Skoog and West (1976), Bockris and Reddy (1977), Bard and Faulkner (1980), Tietz (1986), Christian and O'Reilly (1986). For more details the reader is referred to these standard works.

2.1   ELECTRODE PROCESSES

The processes at an electrode in an electrolyte are very complex. A large variety of parameters influence the rate of the electrode reaction. In figure 2.1 the most important variables are depicted; one can distinguish electrode, mass transfer, electrolyte, electrical and external variables. All these variables affect the electrode reaction. The aim of electrochemical theories is to derive a model for this complex process. Firstly, some definitions are needed for the description of electrochemical processes; thereafter the thermodynamics of the electrochemical cell and the kinetics of the electrode reactions will be discussed. In this section the Nernst and Butler–Volmer equation will be detailed. To conclude, the different models for the electrochemical double layer will be described.

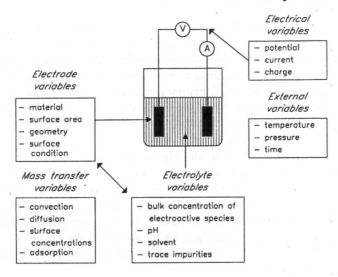

Figure 2.1 An overview of the different variables affecting the electrochemical reaction (adapted from Bard and Faulkner (1980)).

2.1.1 Electrochemical cells

2.1.1.1 Definitions. An *electrochemical cell* consists of at least three components; two electrodes and an electrolyte. The *electrolyte* is a phase through which charge is transferred by the movement of ions. An electrolyte can be a liquid solution, fused salts or ionically conducting solids such as LaF_3. The *electrodes* can consist of metals or semiconductors. They can be solid or liquid. The electrode is a phase through which charge is transferred by electronic movement. Electrochemistry deals with the processes that transport the charge across interfaces between the electrode phase and the electrolyte phase. This transport through the interface involves a change from ionic conduction to electronic conduction.

When two different electrodes are placed in an electrolyte, a potential difference appears and can be measured between the electrodes. This external *cell potential* is a sum of the differences in electrical potential between the various phases in the current path. The measurement and control of this cell potential and cell current is the main topic of experimental electrochemistry.

One would like to study only one electrode–electrolyte interface at a time but one cannot deal experimentally with such an isolated boundary, called a *half-cell*. The reaction taking place in a half-cell is named a *half-reaction*. The overall reaction taking place in a cell with two electrodes is then made up of two independent half-reactions. One is mainly interested in one half-reaction and the potential over one interface. The electrode at which this reaction of interest occurs is named the *working electrode* (WE) or indicator electrode. An electrode at which the half-cell potential is constant is used for the second half-reaction. This electrode, made up of phases with constant composition, is named the reference electrode (RE). The internationally accepted primary reference is the *standard hydrogen electrode* (SHE) or *normal hydrogen electrode* (NHE). This electrode has all components at unit activity. It consists of a Pt wire in a stream of hydrogen at 1 atm. and in HCl (1 N).

If a potential is applied to the electrodes, the difference between this potential and the external cell potential in equilibrium is named the *over-potential*. The potential is always measured at the working electrode with respect to the reference electrode. By applying a more negative potential, the electrode energy is increased. At a level high enough to occupy vacant states on species in the electrolyte, electrons flow from the electrode to the solution. A *reduction current* is observed through the working electrode-electrolyte interface.

By applying a more positive potential, the energy of the electrons is lowered. Electrons can in this case eventually transfer to the electrode where they will find a more favourable energy level. This electron flow from electrolyte to the working electrode is an *oxidation current*. The electrode at which an oxidation current flows is the *anode*. The electrode at which a reduction current is generated is the *cathode*. The critical potential at which these electrochemical reactions can occur is related to the *standard potential E^0*, which is defined by the electroactive species in the solution.

Two different types of electrochemical cells can be distinguished: galvanic and electrolytic cells. A typical example, adapted from Bard and Faulkner (1980), is shown in figure 2.2. In a *galvanic cell* reactions occur spontaneously at the electrodes when connected by a conductor. In an *electrolytic cell* reactions occur only by applying an external voltage to the electrodes. It is worth noticing that for a galvanic cell the anode has a negative potential and for an electrolytic cell it has a positive potential. So, the terms anodic and cathodic refer to electron flow or current direction, not to potentials. In this work mainly electrolytic cells are discussed.

Reversibility is also a term often used in electrochemistry. Unfortunately the terms 'reversibility' and 'reversible' have three different meanings in electrochemical literature. A process is said to be *chemically reversible* if, by reversing the potential at the electrode, the chemical reaction also reverses. So, by reversing the cell current the cell reaction is also reversed. For example, a process is said to be chemically irreversible if, by reversing the cell potential, the chemical process is not reversed or does not take place (e.g. due to side reaction phenomena)

A process is *thermodynamically reversible* when an infinitessimal reversal in a driving force causes the process to reverse its direction. In this case the system must essentially always be at equilibrium. If two states of

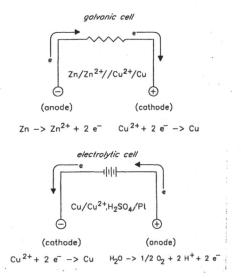

Figure 2.2   The difference between a galvanic and an electrolytic cell (adapted from Bard and Faulkner (1980)).

the system are connected through a continuous series of equilibrium states, a thermodynamicallyreversible path exists between those two states. It would take an infinite length of time to traverse this path. Thermodynamically, reversibility is therefore a theoretical notion as all practical processes occur at a final rate.

A third meaning is *practical reversibility*. A process is said to be practically reversible if the electrode system follows the Nernst equation or an equation derived from it. However, this definition is not absolute; it depends on the experimental conditions and the expectations of the observer. In this work the term 'reversible' is mostly used according to this definition.

In figure 2.3 the convention for plotting current–potential curves, or voltammograms, is shown as used in this book. The potential is plotted in abscissa in an increasing positive sense from right to left. Cathodic currents corresponding to a reduction process are plotted upwards in ordinate. Reduction currents are positive by definition. This implies that anodic currents corresponding to an oxidation process are plotted downwards. All voltammograms shown in this work are plotted according to this convention.

Different measuring techniques can be distinguished. For *potentiometry* the equilibrium potential is measured without current flow through the electrochemical cell. It is a direct analytical application of the Nernst equation. *Voltammetry* can be defined as a measuring technique where the current through the cell is recorded as a function of the applied potential. According to the potential waveform applied to the cell, different voltam-

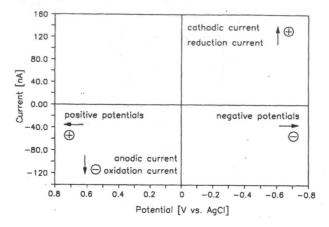

Figure 2.3   The convention for plotting current–potential curves (or voltammograms) used in this book.

metric techniques can be defined: e.g. cyclic voltammetry, linear-sweep voltammetry and pulse voltammetry. *Amperometry* is a special form of voltammetry; the potential is kept constant and the current is recorded as a function of time. *Conductometry* can be defined as a measurement of the conductance of the electrolyte by applying a small-amplitude AC voltage and measuring the AC current. These different techniques are discussed in detail in chapter 3.

2.1.1.2   Thermodynamics of the electrochemical cell.   In this section an equation for the potential at a half-cell is derived. This equation is known as the Nernst equation. This deduction of the Nernst equation is based on electrochemical thermodynamics.

For the study of a half-cell a reference electrode is necessary. Consider a general cell with, on the left-hand side, a NHE with all components at unit activity. At this electrode the following reaction takes place:

$$2H^+ + 2e^- \rightleftharpoons H_2. \tag{2.1}$$

The right-hand half-cell is a general O/R system:

$$aO + ne^- \rightleftharpoons bR \tag{2.2}$$

So, the overall cell reaction is then

$$(n/2)H_2 + aO \rightleftharpoons bR + nH^+. \tag{2.3}$$

The *Gibbs free energy change* of this cell reaction in equilibrium is given from basic thermodynamics by

$$DG = DG_0 + RT \ln \frac{(R)^b (H^+)^n}{(O)^a (H_2)^{n/2}}. \tag{2.4}$$

The quantities between parenthesis represent *activities.*

It can also be derived that (Bard and Faulkner 1980):

$$\Delta G = -nFE \tag{2.5}$$

and,

$$\Delta G_0 = -nFE^0. \tag{2.6}$$

Or, substituting (2.5) and (2.6) into (2.4),

$$E = E^0 - \frac{RT}{nF}\ln\frac{(R)^b(\mathrm{H}^+)^n}{(O)^a(\mathrm{H}_2)^{n/2}}. \tag{2.7}$$

Since in the NHE all components have unity activity, $(\mathrm{H}^+) = 1 = (\mathrm{H}_2)$, equation (2.7) can be simplified to

$$E = E^0 + \frac{RT}{nF}\ln\frac{(O)^a}{(R)^b}. \tag{2.8}$$

This relation, known as the Nernst equation, gives the potential of a general O/R electrode reaction as a function of the activities of O and R for a system in equilibrium. The potential is referred to the NHE. The EMF of the electrochemical cell can now be defined as the difference in potential of the two half-reactions, E_cathode and E_anode:

$$E_\text{cell} = E_\text{cathode} - E_\text{anode}. \tag{2.9}$$

Equation (2.8) is not practical in experimental electrochemistry, as the activity of the electroactive species is usually not known. Equation (2.8) can be converted into concentrations if, instead of the standard electrode potential E^0, the *formal potential $E^{0'}$* is introduced (Bard and Faulkner 1980). This quantity is a measured potential and it incorporates the standard potential and several activity coefficients. The Nernst relation can then be written in a more practical form.

$$E = E^{0'} + \frac{RT}{nF}\ln\frac{[O]^a}{[R]^b}. \tag{2.10}$$

The quantities between brackets now represent the bulk concentrations. The formal potential however varies from medium to medium and is dependent of the ionic strength of the electrolyte. This equation is the basis of the analytical technique known as potentiometry. As can be seen in formula (2.10), the potential varies according to the logarithm of the concentration.

2.1.1.3 Kinetics of the electrode reactions. In the theory of electrode
kinetics, the current through an electrochemical cell is studied as a function
of the potential. In 1905 Tafel derived an experimental law that stipulated
that the current is usually related exponentially to the overpotential η.
Tafel formulated his law as

$$\eta = a + b \log i. \tag{2.11}$$

This relation is sometimes written in an exponential form as

$$i = a'e^{\eta/b'}. \tag{2.12}$$

A successful model of electrode kinetics must be compatible with the Tafel
equation and the Nernst equation. Such a model was derived by Butler
and Volmer. This equation gives a relation between the rate constant and
the overpotential. Via the reaction rate this equation can also be related
to the cell current.

Consider the following reaction:

$$O + ne^- \underset{k_b}{\overset{k_f}{\rightleftarrows}} R \tag{2.13}$$

in which k_f and k_b are respectively the forward and backward *rate constants*. The *forward reaction rate* v_f can be described as

$$v_f = k_f C_O(0,t) = \frac{i_c}{nFA}. \tag{2.14}$$

In this equation $C_O(0,t)$ is the surface concentration of the oxidized form
in the standard system, i_c is the cathodic current and A is the electrode
area.

A similar equation can be derived for the *backward reaction rate* v_b:

$$v_b = k_b C_R(0,t) = \frac{i_a}{nFA}. \tag{2.15}$$

The net reaction rate v_{net} is

$$v_{net} = v_f - v_b = k_f C_O(0,t) - k_b C_R(0,t) = \frac{i_a}{nFA} \tag{2.16}$$

or

$$i = i_c - i_a = nFA(k_f C_O(0,t) - k_b C_R(0,t)). \tag{2.17}$$

Equation (2.17) does not give any information about the applied potential.
The link between equation (2.17) and the applied potential is given by the
Butler–Volmer formulation of electrode kinetics (formula (2.18)). These
equations can be derived from a model based on free-energy curves or a

kinetic model based on electrochemical potentials. Both models result in the same equations (Bard and Faulkner 1980):

$$k_{\mathrm{f}} = k^0 \exp(-\alpha n f(E - E^{0'})) \tag{2.18a}$$

$$k_{\mathrm{b}} = k^0 \exp((1 - \alpha)n f(E - E^{0'})) \tag{2.18b}$$

where $f = F/RT$ (in units of V^{-1}), k^0 is the standard rate constant (in cm s^{-1}), and α is the transfer coefficient.

Substituting (2.18) into (2.17) gives the *complete current–potential characteristic*:

$$
\begin{aligned}
i = nFAk^0[&C_{\mathrm{O}}(0,t) \exp(-\alpha n f(E - E^{0'}) \\
&- C_{\mathrm{R}}(0,t) \exp((1 - \alpha)n f(E - E^{0'})].
\end{aligned}
\tag{2.19}
$$

This relation is compatible with the Nernst equation. At equilibrium the net current is zero and the concentration at the surface equals the bulk concentration. So, equation (2.19) simplifies to

$$\exp\left(\frac{nF}{RT}(E - E^{0'})\right) = \frac{[\mathrm{O}]}{[\mathrm{R}]}. \tag{2.20}$$

Equation (2.20) is in fact an exponential form of the Nernst equation (2.10). Also, under certain conditions equation (2.19) simplifies to the Tafel equation (2.12) (Bard and Faulkner 1980). This proves that the complete current–potential characteristic also covers the more simple cases. This equation is therefore the foundation of the treatment of all electrochemical problems.

2.1.2 The electrochemical double layer

At the interface between electrode and electrolyte an electrochemical double layer is built up. A model for the space-charge region surrounding the electrode is discussed in this section. Helmholtz proposed that the charge distribution in the electrolyte is equivalent to the charge distribution in a metallic electrode, i.e. all excess charge resides strictly at the surface. He modelled the structure of the double layer as equivalent to a parallel-plate capacitor with a plate separation in the molecular order (figure 2.4). The *double-layer capacitance* C_{d} is therefore :

$$C_{\mathrm{d}} = \frac{\varepsilon_{\mathrm{r}}\varepsilon_0}{d} \tag{2.21}$$

where ε_{r} is the relative dielectric constant of the electrolyte, ε_0 is the permittivity of free space ($= 8.86 \times 10^{-12}$ C V^{-1} m^{-1}) and d is the plate separation.

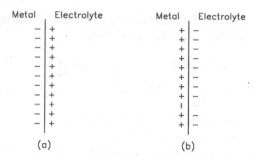

Figure 2.4   The Helmholtz model for the electrochemical double layer.

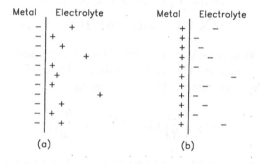

Figure 2.5   The Goüy and Chapman model for the electrochemical double layer.

However, this model is not perfect as, according to (2.21), the double-layer capacitance is a constant. In practice, C_d depends on concentrations and potential.

Goüy and Chapman independently introduced the idea of a *diffuse layer* where the highest concentration of excess charge is found near the electrode surface (see figure 2.5). It is important to notice that the diffuse layer is not identical to the diffusion layer discussed further in this book. The diffuse layer describes the behaviour of the excess charge at an electrode. The diffusion layer can be defined as the region surrounding a current-conducting electrode where the concentration of O or R differs from the bulk concentration.

In the diffuse layer, the electrostatic forces are strong enough to overcome the thermal agitation processes. At further distances the concentration of excess charge drops as the electrostatic forces weaken. With this model, dependence of concentrations and potential can be accounted for. At high potentials a high excess charge builds up at the electrode surface and the diffuse layer becomes more compact. A rise in concentration will also result in a compression of the diffuse layer. A compression of the diffuse layer will result in an increase in the capacitance. This corresponds well with

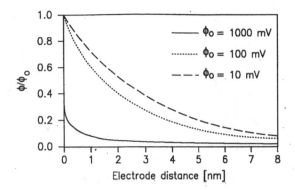

Figure 2.6   The potential profiles through the diffuse layer according to the Goüy and Chapman model.

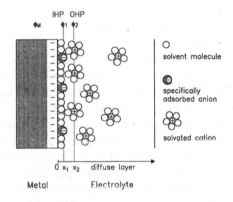

Figure 2.7   The contemporary accepted model of the electrode–electrolyte double-layer region as derived by Stern and Graham (adapted from Bard and Faulkner (1980)).

experimental data. The diffuse layer is relatively thin in comparison to the thickness of the diffusion layer encountered in faradaic experiments. For concentrations greater than 10^{-2} M, the thickness of the diffuse layer is less than 30 nm. In figure 2.6 the potential profiles through the diffuse layer are shown. An exponential form is found for small values of the potential drop φ_0 across the diffuse layer. The potential drop near the electrode becomes steeper for higher potentials.

The presently accepted model of the electrode–electrolyte double-layer region is depicted in figure 2.7; it was derived by Stern and Graham. The electrolyte adjacent to the electrode is thought to be made up of several layers. The layer near the electrode (the inner layer) contains solvent molecules and *specifically adsorbed ions*. This inner layer is called the compact, Helmholtz or Stern layer. The plane of the electrical centres of the specifically adsorbed ions is named the *inner Helmholtz plane* (IHP).

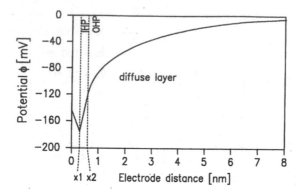

Figure 2.8   The potential profile according to the Stern and Graham model.

This plane is at a distance x_1 from the electrode.

Solvated ions can only approach the electrode to a distance x_2. This distance can be as close to the electrode as 0.5 nm. In the previous model, ions could approach the electrode to a distance equal to zero; in that model the ions were thought to be point charges. The plane of the centres of the solvated ions is named the *outer Helmholtz plane* (OHP). Since only long-range electrostatic forces influence the interaction of the solvated ions with the electrode, these ions are said to be *non-specifically adsorbed*. Because of thermal agitation, the non-specifically adsorbed ions are distributed around the electrode in a diffuse layer. The potential profile through the solution is depicted in figure 2.8. Beside the exponential profile in the diffuse layer, a linear profile is found between the surface of the electrode and the OHP. The total double layer can now be seen as the serial connection of two capacitances. The first capacitance corresponds to the charges held at the OHP and is independent of the potential. The second capacitance corresponds to the diffuse charge and is dependent of the potential. The total capacitance is rather complex and is controlled by the smaller one of the two capacitances.

The electrochemical double layer is of great importance for the response of electrochemical sensors. Capacitive as well as adsorption and faradaic currents find their origin in the double layer. This will be discussed in more detail in section 2.3.1

2.2   POTENTIOMETRY

Potentiometry can be defined as a direct analytical application of the Nernst equation through measurement of the potential of nonpolarised electrodes under conditions of zero current. The measurement is thought to be performed in thermodynamic equilibrium. In the early times of electrochemistry these measurements were carried out with a potentiometer, hence the name potentiometry. Nowadays, a digital voltmeter with high input impedance (> 10 GΩ) is used.

Since usually only the potential of the half-reaction is of interest, a reference electrode is necessary. Reference electrodes are therefore of special importance for electrochemical experiments. Without a stable reference electrode, no reliable electrochemical sensor can be made. Reference electrodes are also based on the principles of the Nernst equation. In this section, the NHE, the calomel and the Ag/AgCl electrode are detailed.

In principle a cell consisting of, for example, a reference electrode and a Pt electrode in an electrolyte containing the redox couple (Fe^{2+}/Fe^{3+}) allows the determination of the Fe^{3+} concentration. However, this type of direct application of the Nernst equation is not used for practical analytical measurements. Such a measurement is not selective and the cross sensitivity to other elements in the solution is therefore too large. To obtain a selective response, ion-selective membranes are used. These classes of electrodes are therefore named ion-selective electrodes (ISE) or membrane electrodes. The best known ISE is without doubt the glass pH electrode. Good ISEs can also be made with solid state elements. These types of devices are therefore worthwhile candidates for planarization. As an example, a new multi-channel ion-selective electrode configuration has been realized with on-chip buffer amplifiers (Lauks *et al* 1985). This smart ion-selective electrode array is detailed in section 5.3.1.

2.2.1 The Nernst equation

As discussed in section 2.1.1, the Nernst equation (2.8), (2.10) gives a simple relation between the potential at an electrode with respect to the solution and the concentration of the corresponding electroactive species in the solution. Measurement of the potential at a reversible electrode thus allows the calculation of the activity or concentration of the electroactive species in the solution. A logarithmic dependence of the potential on the concentration is found. At room temperature (25 °C), a decade change in concentration corresponds to a potential change of 59.1 mV for a univalent reaction. Electrodes or electrode reactions that show this logarithmic dependence are said to be Nernstian.

As a consequence, potentiometry is capable of measuring the concentration over a broad range in most cases. A pH electrode can reliably measure the [H^+] or [OH^-] concentration over more than ten decades. In voltammetric experiments the current varies linearly with the concentration. To cover such a broad range with a voltammetric sensor, a device with low drift and residual current is essential. A highly sophisticated current meter with several sensitivity ranges is also required. Because of the linear dependence of the concentration on the current, voltammetric sensors are, however, more accurate.

In biomedical applications, concentrations change typically by only one or two decades. Therefore voltammetric sensors are preferred for biomedical applications, as the precision of these devices is in principle higher. However, not all clinically significant parameters can be measured in a voltammetric way. For the determination of the pH value, no voltammetric techniques are known. Potentiometry is in this case the ideal technique. Potentiometric sensors are also of special interest if a broad range of concentrations has to be measured or if no electroactive components can be consumed during the measurement.

2.2.2 Applications

2.2.2.1 Reference electrodes.   In a typical electrochemical experiment, it is necessary that the half-cell potential of one electrode is known and is constant and insensitive to the composition under study. An electrode that fits this description is named a reference electrode. Since the potential of the reference electrode is known and constant under all circumstances, changes in the external cell potential are only related to changes in the potential at the working electrode. Therefore reference electrodes are of critical importance in the development of potentiometric and also voltammetric sensors.

Three important types of reference electrodes are known: the normal hydrogen electrode (NHE), the calomel electrode and the Ag/AgCl electrode. All reference electrodes are based on the Nernst equation by keeping all the ion activities constant. A constant electrode potential is thus achieved.

The normal hydrogen reference electrode consists of a platinum wire or foil coated with Pt black (this is finely divided platinum) to increase the surface area. This electrode is placed in a solution with constant hydrogen activity. This is realized by bubbling pure hydrogen over the Pt wire. The pressure of the gas is 1 atm. The continuous stream of hydrogen gas provides the solution with a constant hydrogen concentration. The Pt wire takes no part in the reaction and is only necessary for the electron transfer. Under normal conditions, the NHE behaves as a reversible electrode. It can be used as anode or cathode.

By international convention, the potential at a normal hydrogen electrode is defined to be zero at all temperatures. All standard and formal potentials are referred to this standard. However, the normal hydrogen electrode is not very practical for every day use. Secondary reference electrodes which are more convenient to use are the calomel and Ag/AgCl reference electrode.

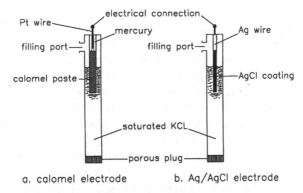

a. calomel electrode   b. Aq/AgCl electrode

Figure 2.9 Schematics of (*a*) the saturated calomel reference electrode and (*b*) the saturated Ag/AgCl reference electrode.

The calomel reference electrode (figure 2.9*a*) consists of a calomel (Hg_2Cl_2) paste electrode immersed in a saturated KCl solution. Electrical contact is made via a Pt wire, contacting the mercury, contacting the calomel paste. Electrochemical contact to the external electrolyte is made via a porous plug or a fritted disk. At 25 °C, the reduction potential versus NHE of the saturated calomel electrode (SCE) is 242 mV. Variants of the saturated calomel electrode exist. Instead of a saturated KCl solution, other KCl concentrations are used. The SCE is available commercially in different forms. It is the most frequently used reference electrode in electrochemical experiments.

The Ag/AgCl reference electrode (figure 2.9*b*) consists of a Ag wire coated with AgCl immersed in a saturated KCl solution. Electrochemical contact to the external electrolyte is made via a porous plug or a fritted disk. The following half-reaction takes place:

$$AgCl(s) + e^- \rightleftarrows Ag(s) + Cl^-. \tag{2.22}$$

The potential of the saturated Ag/AgCl reference electrode is 197 mV with respect to the NHE. A disadvantage of the saturated calomel and the Ag/AgCl reference electrode is the leakage of the internal reference electrolyte into the solution. This leakage prevents the use of this type of electrode for systems where a small change in the halide concentration can not be tolerated.

In biological fluids, such as whole blood or blood plasma, the chloride ion concentration is constant (0.15 M). So, a simple Ag/AgCl wire can in this case be used as a reference electrode, as the potential at this electrode will practically remain constant. Fluctuations in the chloride ion concentration in blood are small; hence, the potential fluctuations will never exceed a few millivolts. For most applications such a pseudo-reference electrode will be satisfactory. In voltammetric experiments the current is usually recorded at a current plateau. Under these circumstances, small deviations of the potential at the reference electrode do not influence the response of a voltammetric sensor. This pseudo-reference electrode (a Ag/AgCl wire or electrode in physiological solution) is in fact not a real reference. In this work, however, it is referred to as a Ag/AgCl reference electrode. The potential of this pseudo Ag/AgCl reference electrode is 272 mV with respect to the NHE. Calomel electrodes are not suited for medical 'in vivo' applications due to the toxicity of mercury.

Ag/AgCl reference electrodes can also easily be prepared using planar techniques. The calomel type is not suited for planar sensors because of the liquid state of mercury. Alternatives, such as the Pd/H_2 electrode, are possible candidates for planar reference electrodes, but their behaviour is not well understood (Tseung and Goffe 1978). Therefore, in this work on planar voltammetric sensors, Ag/AgCl pseudo-reference electrodes are always used.

2.2.2.2 pH electrodes. The best known and mostly used electrochemical sensor is without doubt the glass pH electrode. This sensor is a membrane electrode. The glass membrane is pH sensitive and a potential proportional to the pH value is built up over the glass membrane. The glass membrane allows transport of H^+ ions and forms a barrier for other ions. In figure 2.10 a typical set-up for pH measurements is shown. For the measurement of pH, two electrodes are necessary; a pH-sensitive glass electrode and a reference electrode. In modern pH electrodes the two electrodes are combined into one glass or plastic tube so that the external appearance corresponds to a single electrode.

The pH glass electrode consists of a Ag wire coated with AgCl. This wire is immersed in a 0.1 M HCl reference solution. A thin pH-sensitive glass bulb is sealed on the tip of the electrode. The total configuration can be thought of as two reference electrodes, the potential of which is independent of pH, separated by a pH-sensitive glass membrane.

If the external potential of this cell is measured as a function of the internal and external hydrogen ion activities a_{in} and a_{ext}, a relation identical to the Nernst equation is found:

$$E = E_c + \frac{RT}{nF} \ln \frac{a_{ext}}{a_{in}}. \tag{2.23}$$

The value of E_c is a constant consisting of the potential at the SCE, the Ag/AgCl electrode and the asymmetry potential. The causes of the asymmetry potential are obscure and its effect is eliminated by calibrating the pH electrode regularly.

At 25 °C, and taking into account that the internal hydrogen ion activity is constant, equation (2.23) can be simplified to :

$$E = K + 0.0591 \log a_{ext} \tag{2.24}$$

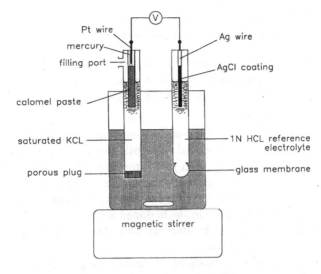

Figure 2.10   A typical set-up for pH measurement.

or

$$E = K - 0.0591\,\mathrm{pH} \qquad (2.25)$$

where K is a new constant containing a_{in}.

A typical glass membrane which ranges in thickness between 30 mm and 100 mm has the following composition: 22% Na_2O, 6% CaO and 72% SiO_2. It is known that the surface of the glass membrane must be hydrated in order to function as a pH electrode. A gel is formed at the surface with a thickness from 0.01 to 0.1 μm. The current conduction in the membrane is ionic and involves the movement of Na^+ ions in the glass membrane. In the gel layer the current is conducted by Na^+ ions and H^+ ions with an increasing proportion of conducting H^+ ions at the outer surface of the gel layer. At the gel–electrolyte interface, current conduction involves the transfer of protons. These processes are the source of the potential over the glass sensitive membrane. The impedance of the glass membrane can be as high as 100 $M\Omega$.

Unfortunately, the concept of a glass pH electrode cannot be converted into a planar mass-produced pH electrode, since an internal reference electrolyte is necessary. Only one attempt to realize a planar glass pH electrode is reported in the literature (Belford *et al* 1987).

There are, however, other principles to measure the pH. A complete solid-state alternative is observed by noble metal oxide electrodes (Fog and Buck 1984). These solid-state membrane ion-selective electrodes, such as IrO_2 and PtO, also show a near Nernstian response to a varying pH. In particular, the IrO_2 electrode has been studied extensively in the literature. This type of pH electrode has also been realized using planar techniques

(Lauks *et al* 1985). Another planar approach for pH electrodes is the ISFET. This device is discussed more in detail in section 2.5.

2.2.2.3 Ion-selective electrodes. Although the glass pH electrode is also an ion-selective membrane electrode, it is usually classified in a separate category. Three other categories can be distinguished:

1. Solid-state membranes

These ISEs are based on solid-state membrane electrolytes which have tendencies toward the preferential adsorption of certain ions on their surface. The glass pH electrode is in fact a member of this category.

The best known electrode in this category is without doubt the single-crystal LaF_3 membrane electrode used for the determination of fluoride ion concentrations. This membrane is doped with EuF_2 to allow ionic conduction through the crystal by fluoride ions. In this way, the electrode is selective towards fluoride ions.

Other examples of this category are based on precipitates of insoluble salts, such as silver halides. The surface of these membranes is sensitive to the ions included in the salt and to other species that form more insoluble precipitates.

2. Liquid membranes

Liquid-membrane electrodes are based on the potential that is built up across the interface between the analyte and an immiscible liquid that selectively bonds with the ion being determined. Instead of the solid membrane separating the solution of fixed and unknown activity, now a thin layer of an immiscible organic liquid is used. As shown in figure 2.11, a porous organophilic hydrophobic membrane serves to hold the organic layer between the two aqueous solutions. The organic layer contains chelating agents with selectivity towards the ion of interest. Commercial calcium-selective electrodes are based on this principle.

However, the principle of liquid membranes can not be used for the realization of planar sensors due to the liquid state of the essential component.

3. Neutral carriers

The ion-selective membrane electrodes described above are based on the flux of charged particles for the transport of the charge from one side of the membrane to the other. The neutral-carrier membranes are based on uncharged chelating agents: the neutral carriers. The transport of charge is enabled by selectively complexing the ion of interest. An important commercial device based on this principle is the potassium-selective electrode with the macrocycle valinomycin as neutral carrier in a PVC matrix. Valinomycin is a natural antibiotic with the structure shown in figure 2.12. Inside this molecule is a hole where the complexing reaction takes place. The selective characteristics of this device rely on the fit between the size of the ion to be complexed and the volume of the complexing cavity in the macrocycle.

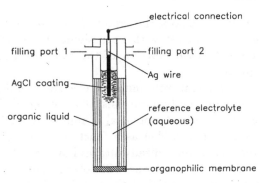

Figure 2.11 The cross section of a liquid membrane ion-selective electrode.

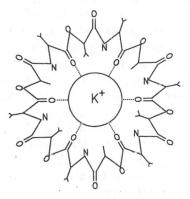

Figure 2.12 The two-dimensional chemical structure of the valinomycin $(C_{54}H_{90}O_{18}N_6)$-potassium complex.

A lot more can be said on ion-selective electrodes, especially about the solid-state and the neutral-carrier membrane electrodes, since they can be realized as planar devices with microelectronic techniques. A detailed study of this type of electrode, however, is beyond the scope of this work. The reader is therefore referred to the abundant literature on this topic: e.g. Freiser (1978, 1980), Solsky (1988) and previous review papers in Analytical Chemistry.

2.3 VOLTAMMETRY

Voltammetry is the measurement of the current which flows at an electrode as a function of the potential applied to the electrode. As a result of a voltammetric experiment, the current–potential curve, or voltammogram, is recorded. These curves can be used for qualitative and quantitative determinations and for thermodynamic and kinetic studies. As this book deals with electrochemical sensors, the aspect of quantitative determination is most important. In contrast with potentiometry, voltammetry gives a linear current response as a function of the concentration, which is a considerable advantage.

In figure 2.13 two different voltammetric electrode configurations are shown. The two-electrode system consists of a reference electrode (RE) and a working electrode (WE). A potential is applied to this electrochemical cell and the current is recorded as a function of this potential. The two-electrode system has several disadvantages however. As the reference electrode carries current, the electrode will polarize and an overpotential will occur. Therefore, the potential at the working electrode is unknown. This polarization can be avoided by the use of a very large reference electrode and a small working electrode. In this way, the current density will be low enough to prevent polarization of the electrode.

Material consumption in the reference electrode is also a problem encountered in a two-electrode system. A good example of this problem is an oxygen sensor consisting of a Au working electrode and a Ag/AgCl reference electrode. To reduce the oxygen at the cathode (WE), an oxidation of Ag to AgCl will take place at the anode (RE). If all Ag in the reference electrode is consumed, the reference electrode will not function properly and the potential at the working electrode will be unknown. This problem can also be avoided by the use of a very large reference electrode and a small working electrode. Large electrodes are no problem in laboratory experiments. For biomedical applications the size of the electrodes is critical, as 'in vivo' use is the ultimate objective.

A better approach is the use of a three-electrode system in a potentiostatic configuration. Beside the working and reference electrode, an auxiliary (AE) or counter electrode (CE) is introduced (see figure 2.13). The reference electrode in the two electrode system is split up in two-electrodes: a true reference electrode for controlling the potential and an auxiliary electrode for current injection in the electrolyte. In a three-electrode voltammetric system a potentiostat controls the current at the auxiliary electrode as a function of the potential. This is realized in practice with an operational amplifier (opamp) as shown in figure 2.13.

The operation can be explained as follows. The potential is applied to the positive input of the opamp. The reference electrode is connected to the negative input and measures the potential in the solution. The auxiliary electrode is connected to the output. The opamp injects a current into the solution through the auxiliary electrode. Due to the feedback mechanism of the opamp, the current is controlled in such a way that the potential at the negative input (RE) equals the potential at the positive input. The potential difference between the working electrode and the reference electrode equals the applied potential. As no current is flowing through the reference electrode (the opamp has a very high input impedance), this potential is

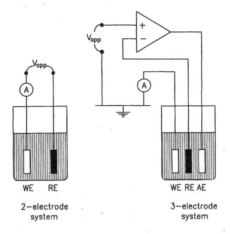

Figure 2.13 The two-electrode and three-electrode voltammetric electrode configuration.

well defined and not influenced by the current density. The current is measured at the working electrode. This current is in principle identical to the current through the auxiliary electrode if no parasitic current paths exist. As will be discussed in section 2.3.1 this current is linearly proportional to the concentration of the species involved and selectivity is obtained by the proper choice of potential and or by use of a selective membrane.

The current at the working electrode is the sum of three different current-inducing processes; one can distinguish capacitive currents, faradaic currents and absorption effects. These processes result in a large variety of possible responses of the sensor to a varying potential. In the next section the theoretical background of the following measurements will be discussed.

1. Linear-sweep voltammogram of a Au electrode in PBS saturated with air and with nitrogen (figure 2.14)

A gradually increasing potential from -200 mV to -900 mV is applied to a Au electrode. If the solution is saturated with nitrogen, a current practically equal to zero is recorded. If dissolved oxygen is present in the solution, a steep current rise is observed starting at -300 mV. A further increase in potential results in stabilization of the current. The value of the limiting current is linearly proportional to the dissolved oxygen concentration.

2. Cyclic voltammogram of a RuO$_2$ electrode in PBS with [Fe(CN)$_6$]$^{4-}$ (figure 2.15)

A triangular potential wave is applied to a RuO$_2$ electrode. In pure PBS a rectangular voltammogram is obtained. If [Fe(CN)$_6$]$^{4-}$ is added to the solution, two peaks followed by a current plateau appear in the voltammogram. The current at these peaks and the plateaus is linearly related to the [Fe(CN)$_6$]$^{4-}$ concentration.

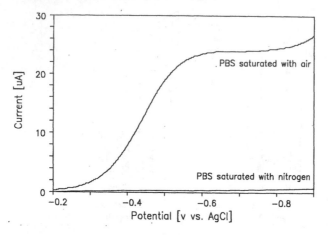

Figure 2.14   Linear-sweep voltammogram of a Au electrode in PBS saturated with air and with nitrogen.

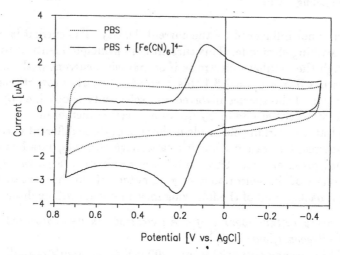

Figure 2.15   Cyclic voltammogram of a RuO_2 electrode in PBS with $[Fe(CN)_6]^{4-}$.

3. Cyclic voltammogram of a Pt electrode in 1 M H_2SO_4 (figure 2.16)

A triangular potential wave is applied to a Pt electrode in a 1M H_2SO_4 solution. Although no intentionally added electroactive species are present in the solution, a complex combination of peaks and plateaus is observed. The shape of the voltammogram is typical for a Pt electrode.

These examples are taken from experiments performed on microelectrochemical sensors described in this book. These voltammetric measurements will be explained in the next section with the help of basic electrochemical theory.

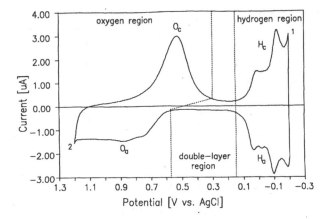

Figure 2.16   Cyclic voltammogram of a Pt electrode in 1 M H_2SO_4.

2.3.1   Theory

The current measured during a voltammetric experiment has three different sources. First, due to the charging and discharging of the electrochemical double layer, a capacitive current is flowing through the cell. Second, the faradaic current, which finds its origin in electrochemical reactions, produces a change in redox state of some solution species or the electrode material itself. Third, an electrode immersed in an electrolyte also experiences adsorption effects.

Double-layer charging and adsorption contribute to the background or residual current. Another part of the residual current finds its origin in the electrochemical reaction of trace impurities in the solution or at the electrode surface. In electrochemical experiments much attention is directed to the minimization of the residual current. The faradaic current is the current of interest in voltammetric analysis.

2.3.1.1 Capacitive currents.   As discussed in section 2.1.2, an electrode immersed in an electrolyte is surrounded by a space-charge region. This region is named the electrochemical double layer and is characterized by the double-layer capacitance C_d. This capacitance depends on the electrolyte composition and the applied potential. If a varying potential as a function of time is applied to an electrode, this layer has to be charged and discharged. This charging process gives rise to a capacitive current i_c

$$i_c = C_d \frac{dv}{dt}. \tag{2.26}$$

If a triangular potential wave is applied to a purely capacitive cell, the current through the cell can be defined as

$$i_c = C_d S \tag{2.27}$$

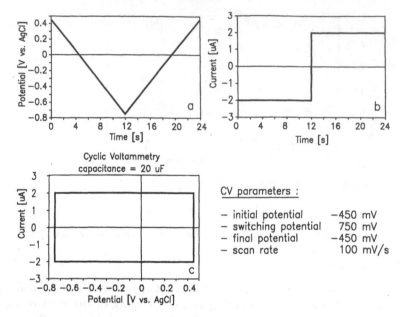

Figure 2.17 Applied potential, measured current and cyclic voltammogram of a purely capacitive cell (capacitance = 20 μF).

where S equals the scan rate

$$S = \frac{dv}{dt}. \tag{2.28}$$

The scan rate is usually expressed in (mV sec^{-1}) and equals the slope of the triangular potential wave.

The resulting capacitive current is sketched in figure 2.17. The current–potential curve is in this case a rectangle. It is important to notice that the current through a capacitive cell is proportional to the scan rate. This corresponds well with the observations made in figure 2.15. In PBS the RuO$_2$ electrode behaves like a capacitor; the voltammogram is practically a rectangle if no electroactive species are present in the electrolyte. The voltammogram for the RuO$_2$ electrode is not a perfect rectangle; this is related to the potential dependence of the capacitance on the applied potential and to the presence of resistive elements.

The capacitive current does not give valuable analytical information and is therefore an interfering current in analytical measurements. It is usually the largest component of the residual current; i.e. the current measured without intentionally added electroactive species in the solution. The other components of the residual current are related to adsorption effects and the current due to oxidation or reduction of trace impurities in the electrolyte. In laboratory experiments this last current can be kept small by use of pure materials and a clean environment. The other components of the residual current can be kept small by decreasing the scan rate.

2.3.1.2 Adsorption effects.

2.3.1.2 Adsorption effects. If the cyclic voltammogram of a Pt electrode is taken in 1M H_2SO_4, a similar result is expected as for a RuO_2 electrode in PBS, as no electroactive species are present in the solution. However, instead of a rectangular shape, the voltammogram shows a large variety of peaks. The peaks are due to the oxidation and reduction of Pt at the surface of the electrode and to the inherent reactivity of this electrode material towards hydrogen oxidation and reduction.

The following peaks are observed on the Pt voltammogram (figure 2.16):

O_a formation of adsorbed oxygen and platinum oxide layers
O_c reduction of oxide layers
H_a oxidation of adsorbed hydrogen (several peaks)
H_c reduction of adsorbed hydrogen (several peaks)

The capacitive current described in the previous section is only observed in the double-layer region. The peak current, i_a, due to adsorption effects, is given by Bard and Faulkner (1980):

$$i_a = \frac{n^2 F^2}{4RT} A \Gamma_O^* S. \tag{2.29}$$

In this equation Γ_O^* is the amount of the species adsorbed on the surface of the electrode per unit area. It is important to notice that the peak current and the current at all the other points on the adsorption peaks in figure 2.16 are proportional to the scan rate S. An identical relation is observed for the capacitive current.

Although the redox state of the adsorbed material at the electrode can change and hence a current can result, this current has different characteristics than the real faradaic current described in the following section. The most important difference is that diffusion does not play an important role in adsorption effects.

The shape of the adsorption peaks and waves gives valuable information on electrode contamination and electrode quality. The charge under the hydrogen wave can also be used to calculate the chemically active electrode area of Pt electrodes. Therefore, cyclic voltammograms of Pt electrodes can be used for the evaluation of the surface state of the electrode.

Adsorption phenomena are very sensitive to the electrode surface and so to the history of the electrode. To obtain a constant response, some electrode materials need pre-treatment. In particular, Pt electrodes are known for their unpredictable behaviour with respect to hydrogen adsorption. Therefore it is imperative to adopt some standard set of experimental procedures when using Pt electrodes. For instance, these procedures are detailed in Adams (1969) and involve, for example, chemical cleaning in hot nitric acid. However, these cleaning procedures can not be followed in the biosensor application anticipated in this work; only pre-polarization techniques can be used. Because of the irreproducible behaviour of Pt electrodes in the positive potential domain, these electrodes are not recommended for experiments in this domain.

2.3.1.3 Faradaic currents. The faradaic current is a measure of the rate of the electrochemical reaction taking place at the electrode. Faradaic currents also produce a change in redox state of some material in the cell. This current is directly proportional to the concentration of reduced and oxidized species and is therefore most interesting for analytical applications.

The rate of the electrochemical reaction taking place is determined by two processes; the mass transport and the charge transfer. The mass transport is the process that transports material from the bulk of the solution to the electrode. The charge transfer is the process that transfers electrons from the electrode to the solution and vice versa.

Mass transport can take place by the following modes:

1. *diffusion*—the movement of material from high-concentration regions to low-concentration regions;

2. *migration*—the movement of charged particles in an electric field;

3. *convection*—the movement of material contained within a volume element of a hydrodynamic solution.

The three modes of mass transport are depicted in figure 2.18. The material flux to the electrode is described by the Nernst–Planck equation:

$$J(x,t) = -D\overbrace{\frac{\delta C(x,t)}{\delta x}}^{\text{diffusion}} - \overbrace{\frac{zF}{RT}DC(x,t)\frac{\delta\phi(x,t)}{\delta x}}^{\text{migration}} + \underbrace{C(x,t)v_x(x,t)}_{\text{convection}} \tag{2.30}$$

where F is the Faraday constant (C), J is the material flux (mol cm^{-2} sec^{-1}), D is the diffusion coefficient (cm^2 sec^{-1}), C is the concentration (M) and v_x is the hydrodynamic velocity (cm sec^{-1}).

If only diffusion contributes to the flux at the electrode surface, then the current can be defined as:

$$i(t) = nFAD\frac{\delta C(x,t)}{\delta x}\bigg|_{x=0} \tag{2.31}$$

This means that the current is proportional to the slope of the concentration profile at the surface of the electrode. Since only the diffusion mode

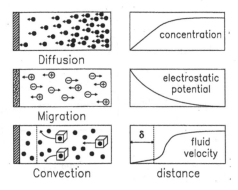

Figure 2.18 The three modes of mass transport (adapted from Maloy (1983)).

is taken into account, it is assumed that no convection exists in the solution or that the solution is not stirred. Also migration is neglected. In a practical experiment migration of the electroactive species is prevented by addition of an inert salt in a 100-fold excess to the solution. By use of this supporting electrolyte, the electric field is dissipated over all the ions in the solution and not solely over the electroactive species. Most of the applications anticipated in this work make use of PBS or physiological solution as electrolyte. Because of the high salt concentrations in these solutions, the migration mode of mass transfer can be excluded.

From (2.30) it can be concluded that if the concentration profile in the vicinity of the electrode surface is known, then the faradaic current can be calculated. For reversible systems the concentration is also determined by the Nernst equation, if the thermodynamic laws are followed. This implies that the charge transfer is sufficiently fast so that an equilibrium exists at every moment. The concentration in the solution also follows Fick's laws of diffusion, hence the current can be determined by solving the partial differential equations. In fact one has to deal with a boundary value problem. The mathematical solution of these problems is beyond the scope of this work and can be found in several text books (Bard and Faulkner 1980, Adams 1969). In this section some solutions of the boundary value problems will be discussed in order to explain the measured current–potential curves. Amperometry, reversible and irreversible steady-state voltammetry and reversible linear-sweep voltammetry will be detailed.

The simplest voltammetric experiment that can be performed is an amperometric experiment. A constant potential as a function of time is applied to the electrode. The potential is sufficiently negative to achieve the limiting current due to the reduction of O to R (reaction (2.13)). The *limiting current* is achieved at the moment that the potential is so negative that the electrode surface concentration equals zero. This happens when the applied electrode potential is more than 118 mV negative with respect to $E^{0'}$ if the

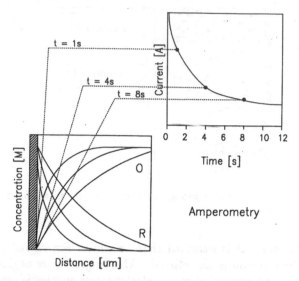

Figure 2.19   The concentration profiles and the current as a function of time during an amperometric experiment.

thermodynamic laws are followed. The current as a function of time can be calculated by solving the boundary value problem. Under semi-infinite linear-diffusion conditions at large plane electrodes, the current is described by the Cottrell equation:

$$i(t) = \frac{nFAD_O^{1/2}C_O^*}{\pi^{1/2}t^{1/2}}. \tag{2.32}$$

In this equation, D_O is the *diffusion coefficient* and C_O^* is the bulk concentration of the oxidized species O. In figure 2.19 the concentration profiles and the current as a function of time are plotted; this representation has been adapted from Maloy (1983). At any moment the applied potential maintains the electrode surface concentration of O at zero. The thickness of the diffusion layer expands as a $t^{1/2}$ function of time. This *diffusion layer* is the region in the vicinity of the electrode where the concentration differs from the bulk concentration. This expansion of the diffusion layer results in a fall-off of the current in proportion to $t^{-1/2}$, since the concentration slope decreases as $t^{-1/2}$. This $t^{-1/2}$ relationship is typical for diffusion phenomena and will also appear in other problems. A further discussion on amperometry can be found in section 3.2.6.

Another basic technique is reversible steady-state voltammetry. In this technique the potential is changed very slowly as a linear function of time. It is assumed that on every point of the voltammogram the Nernst equation is obeyed at the surface of the electrode. The $t^{1/2}$ expansion of the diffusion layer is prevented by stirring the solution so that the thickness of the diffusion layer remains constant during the complete experiment. Although

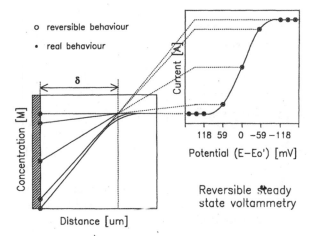

Figure 2.20 A reversible steady-state voltammogram and the corresponding concentration profiles.

the solution is stirred, equation (2.30) is still valid, as the hydrodynamic velocity at the surface of the electrode is zero. The current–potential relationship can then be determined by calculating the concentration at the electrode surface according to the Nernst equation. As the bulk concentration is maintained at a constant distance δ from the electrode surface by stirring, the slope of the concentration profile and hence the current is only determined by the concentration at the electrode surface. The resulting voltammogram is said to be Nernstian. The voltammogram and the corresponding concentration profiles are sketched in figure 2.20.

This voltammogram has some typical characteristics. The limiting current is achieved at potentials more than 118 mV negative with respect to $E^{0'}$ (the potential values are given for experiments at 25 °C). At these potentials, the electrode surface concentration of O is zero. This current i_l is linearly proportional to the bulk concentration C_O^*:

$$i_l = nFAm_O C_O^*. \tag{2.33}$$

The constant m_O is called the *mass transfer coefficient* (in units of m sec^{-1}).

The slope of the voltammogram is also steep. It goes from 10% to 90% of the limiting current over a 118 mV range if n equals unity. This characteristic can be used as a diagnostic bench-mark for a reversible, Nernstian wave.

At a potential equal to $E^{0'}$ the slope of the concentration is half the limiting slope. The current at this potential is then half the limiting current. This specific potential is named the *half-wave potential* and is typical for the O/R system. It can therefore be used for qualitative analysis.

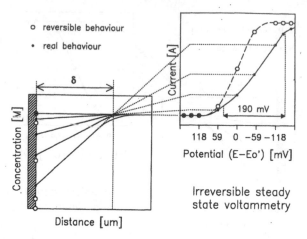

Figure 2.21   An irreversible steady-state voltammogram and the corresponding concentration profiles.

If the system is irreversible, another voltammogram is obtained. In this case the electrode surface concentrations are determined by kinetics instead of the Nernst equation. The behaviour of the system is termed non-Nernstian or irreversible. The resulting voltammogram and the corresponding concentration profiles are sketched in figure 2.21. The electrode surface concentration is determined by the rate of charge transfer and is higher than predicted by thermodynamics. The current at a given potential is thus less than the reversible current. The limiting current is achieved at more negative potentials, as determined by thermodynamics. It is still proportional to the bulk concentration of O and can be used for quantitative analysis; the half-wave potential, however, does not give qualitative information because it depends upon the experimental conditions. It now provides kinetic information about the system.

The voltammogram observed in (2.14) can be explained using the previous discussion about current–potential curves. If no oxygen is present in the solution, only the residual current is observed. This current increases at potentials below −800 mV. At this potential, water starts to dissociate, giving rise to large currents. The use of Au electrodes in PBS is therefore limited to this potential. If oxygen is present in the solution, the following overall reaction takes place:

$$O_2 + 2H_2O + 4e^- \rightarrow 4OH^-. \tag{2.34}$$

It is assumed that this reduction of oxygen is the combination of two reactions, with H_2O_2 as an intermediate reaction product (Hoare 1968):

$$O_2 + 2H^+ + 2e^- \rightarrow H_2O_2 \tag{2.35}$$

and

$$H_2O_2 + 2H^+ + 2e^- \rightarrow 2H_2O. \tag{2.36}$$

A thorough examination of the voltammogram in figure 2.14 shows that the potential difference between the 10% i_l and the 90% i_l point exceeds 220 mV. This implies that the reduction reaction of oxygen is irreversible. Therefore, it can be concluded that the voltammogram in figure 2.14 is completely controlled by kinetics. Nevertheless, it is still valuable for analytical applications since the limiting current is proportional to the bulk dissolved oxygen concentration, even for irreversible systems. The well known Clark cell is based on this principle (see section 2.3.2).

If the scan rate of the applied potential is increased and the voltammetric experiment is done in a quiescent solution so that the diffusion layer increases with time, the resulting response is a reversible linear-sweep voltammogram. The voltammogram and the corresponding concentration profiles are sketched in figure 2.22. In this experiment, Nernstian electrode conditions are assumed. The slope of the concentration profile (and hence the current) changes for two reasons. The electrode surface concentration varies with the potential according to the Nernst equation and the diffusion layer expands according to the $t^{1/2}$ relationship (see Cottrell equation (2.31)). Initially, the process determined by the Nenst equation predominates and the current increases with potential. Thereafter, the diffusion layer expansion process becomes more important and the current decreases. In between, a typical current peak is formed. The value of this peak current peak, i_p, is calculated by solving the boundary-value problem, assuming semi- infinite linear diffusion (Bard and Faulkner 1980):

$$i_p = 0.4463 n F A C_O^* \left[\frac{nF}{RT}\right]^{1/2} S^{1/2} D_O^{1/2}. \tag{2.37}$$

From this formula it can be concluded that a reversible peak current is linearly proportional to the bulk concentration C_O^* and proportional to the square root of the scan rate S. This relation is valid for each point on the voltammogram.

The peak current is attained when the electrode potential is 28.5 mV (25 °C) more negative then $E^{0'}$. Therefore, the value of the peak potential E_p gives valuable qualitative information on the evaluated system. The rising part of the voltammogram is quite steep—it goes from 10% of i_p to i_p over a 100 mV range, if n equals unity. This characteristic can be used as a diagnostic bench mark for a reversible, Nernstian peak.

The voltammogram shown in figure 2.15 can be explained using the above theory. Instead of a single sweep, a triangular potential wave is applied to the electrode. In this way the cathodic and the anodic wave are recorded in one single experiment. In the anodic sweep the $[Fe(CN)_6]^{4-}$ present in

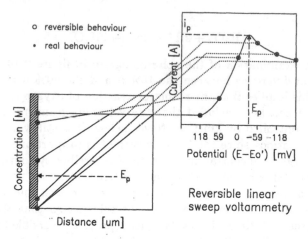

Figure 2.22   A reversible linear-sweep voltammogram and the corresponding concentration profiles.

the solution is oxidized according to the following reaction:

$$[Fe(CN)_6]^{4-} \rightarrow [Fe(CN)_6]^{3-} + e^-. \tag{2.38}$$

The $[Fe(CN)_6]^{3-}$ produced during the anodic sweep is then again reduced according to the opposite reaction:

$$[Fe(CN)_6]^{3-} + e^- \rightarrow [Fe(CN)_6]^{4-}. \tag{2.39}$$

In this way, two peaks are observed on the voltammogram. For a completely reversible system, the peaks are separated by 59 mV from each other. The faradaic current is, however, combined with the capacitive current, resulting in the typical voltammogram shown in figure 2.15. This potential-reversal technique is called cyclic voltammetry.

It can be concluded that the faradaic current results in a large variety of current–potential curves, depending on the applied potential and the reversibility of the electrode reaction. However, it is important to notice that the current is, in practically all cases, linearly proportional to the bulk concentration. Voltammetric sensors can therefore be used advantageously for analytical applications.

2.3.1.4 Conclusion.   The current through an electrochemical cell can have different origins. Each type of current has some typical characteristics summarized in table 2.1. The most important characteristics are the shape of the voltammogram and the dependence on the scan rate. According to these characteristics, the experimental voltammetric response can be explained.

Table 2.1 An overview of the different current types observed in voltammetric experiments.

process	shape	scan rate dependence	typical parameter
ZERO CURRENTS			
capacitive current	rectangular	~ S	C_d
adsorption current	current peak	~ S	i_p
FARADAIC CURRENTS			
amperometry	$t^{-1/2}$ relation	-	$(D/\pi t)^{1/2}$
reversible steady state voltammetry	S-shape	none	$E_{1/2}, i_l$
irreversible steady state voltammetry	S-shape	none	i_l
reversible linear sweep voltammetry	current peak	~ $S^{1/2}$	i_p, E_p

The voltammogram in figure 2.14 (Au electrode in PBS N_2–air) consists of a capacitive current combined with an irreversible steady-state current. The linear-sweep voltammogram in figure 2.15 (RuO$_2$ electrode in PBS $[Fe(CN)_6]^{4-}$) is the combination of a pure capacitive current with a reversible diffusion current. The voltammogram of (2.16) (Pt electrode in H_2SO_4) is the combination of a capacitive current with adsorption effects.

Based on these voltammograms and on the electrochemical principles, all voltammetric experiments in this work can be explained.

2.3.2 Applications

2.3.2.1 The dropping-mercury electrode. The starting point for analytical applications of voltammetric techniques is the dropping-mercury electrode (DME). This electrode was invented by Heyrovsky in the early 1920s. This invention was the start of polarography, now seen as a separate class in voltammetry. For his work on the DME and polarography, Heyrovsky was awarded the 1959 Nobel prize in chemistry.

A typical DME is shown in figure 2.23. A capillary with an internal diameter of approximately 50 μm is fed by a mercury column, typically 20 to 100 cm high. At the tip of the capillary a spherical drop is formed by

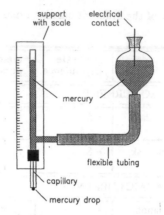

Figure 2.23 Schematic of a dropping-mercury electrode.

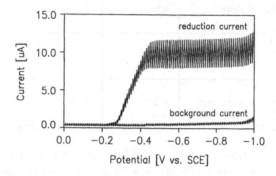

Figure 2.24 Typical polarogram for a DME electrode.

gravity. The drop grows until its weight is no longer supported by surface tension. By increasing the height of the mercury column, the lifetime of the drop is decreased. The maximum diameter of the drops is between 0.5 and 1 mm. The lifetime of a drop is typically 2 to 6 seconds. The continuous flow of identical mercury droplets is used as a working electrode in a potentiostatic configuration.

The DME has several advantages. Mercury has a high hydrogen overvoltage and is thus very interesting for cathodic studies. The electrode surface is continually renewed and hence cannot become fouled or poisoned. Upon falling, the drop stirs the solution so that the next drop is formed into a practically fresh solution. Each drop is in fact a new electrochemical experiment. So, the history of the electrode will not influence the response.

The DME also has disadvantages, however. Mercury is very easily oxidized, so the use of a mercury electrode is restricted to +0.4 V versus SCE for anodic studies. Due to the continuous change in electrode area, current variations corresponding to the drop time are observed in the polarogram (figure 2.24). Sampling or averaging of the current is therefore necessary.

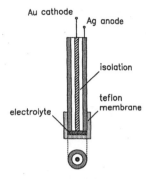

Figure 2.25   Basic configuration of a Clark cell.

The most important disadvantages, however, are the toxicity of mercury and the bulky dimensions of the set-up. These features prevent the use of the DME for 'in vivo' applications. It is also very difficult to imagine a planar version of the DME.

2.3.2.2 The Clark cell.   The Clark cell was discovered by L C Clark in 1956 (Clark 1956) for the measurement of dissolved oxygen concentrations in blood and tissue. The principle of the Clark cell is nowadays extended and finds extensive applications in gaseous oxygen monitors, glucose analysers and instruments used to monitor water quality.

The basic Clark cell consists of two metal electrodes; a cathode and an anode (figure 2.25). The cathode is made in Pt or Au and the anode is made in silver—the area of this electrode is made considerably larger than the area of the cathode. The electrodes are covered by a membrane permeable to oxygen which is separated from the electrodes by an electrolyte. The membrane provides protection of the electrodes against poisoning and electrolyte evaporation and also diminishes the flow dependence.

If a negative potential (-0.7 V versus AgCl) is applied to this cell, then the amount of dissolved oxygen at the cathode (WE) is reduced. The exact electrochemical process is not known. It is assumed that a two-step reaction ((2.34), (2.35)) with H_2O_2 as an intermediate product results in the following global reaction:

$$O_2 + 2H_2O + 4e^- \rightarrow 4OH^-. \tag{2.40}$$

At the Ag anode, Ag is oxidized into AgCl:

$$Ag \rightarrow Ag^+ + e^-. \tag{2.41}$$

The silver ions react with the chloride ions in the electrolyte yielding AgCl. In this way a Ag/AgCl reference electrode is formed so that the potential

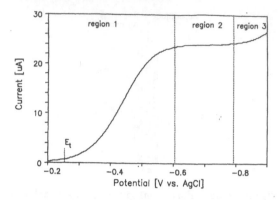

Figure 2.26   A typical response of a Clark cell. Three different regions can be distinguished (scan rate = 20 mV sec^{-1}).

at the cathode is well known. It is important to notice that for each reduced oxygen molecule, four Ag atoms and four Cl atoms are consumed. This implies that a Clark cell has a limited lifetime because of metal and electrolyte consumption at the anode.

In figure 2.26, a typical response of a Clark cell is shown. As discussed in section 2.3.1, an S-shaped voltammogram is obtained. Three different regions can be distinguished.

1. At voltages higher than the threshold voltage E_t, oxygen is reduced at the cathode. This results in a decrease of the electrode surface oxygen concentration. In this region the measured current increases with the applied potential.

2. By increasing the applied potential, limiting current conditions are reached. The electrode surface oxygen concentration equals zero. In this region, the current is controlled by diffusion. A further increase in potential does not result in an increase of the measured current. This limiting current is linearly proportional to the dissolved oxygen concentration.

3. By further increase of the potential, the region is reached where dissociation of water takes place. This results in a steep rise in the current. Higher potentials will destroy the membrane of the Clark cell because of gas evolution.

The limiting current measured with a Clark cell is a linear function of the oxygen concentration and depends on the following parameters (Parker 1987):

$$i_l = \frac{nF}{V} A \frac{p_m}{d_m} pO_2 \qquad (2.42)$$

where V is the molar volume, A is the cathode (WE) area (mm^2), p_m is the oxygen permeability of the membrane (M mm^{-1} Pa^{-1} s^{-1}), d_m is the

Table 2.2   Requirements for the different design parameters of a Clark cell.

parameters-characteristics	electrode area	membrane permeability	membrane thickness
high sensitivity	max	max	min
low flow dependence	min	min	max
fast response time	min	max	min
long lifetime (2-el.)	min	min	max

max = maximised   min = minimised   2-el. = two-electrode system

thickness of the membrane (mm) and pO_2 is the partial oxygen pressure (Pa, mm Hg).

In the design of a Clark cell, all these parameters will influence the response. The most important design parameters are the electrode area, the membrane permeability and the membrane thickness. In table 2.2, the influence of these parameters on sensitivity, flow dependence, response time and lifetime is summarized. For biomedical applications a small cathode area and a thin membrane with a high permeability are frequently cited as a good compromise. However, a small cathode area has the disadvantage that the measured current is low. This can be avoided by the use of multi-cathode devices (Siu and Cobbold 1976).

All Clark cells that are commercially available are based on wire and tube technology. Several catheter-tip devices are described in the literature (Kimmich and Kreuzer 1969, Parker 1987). The cost of these sensors is extremely high because of the manual production of the devices. Clark cells based on planar microelectronic technology are a cheap alternative to these wire and tube devices.

The first planar oxygen sensors are described by Eden *et al* (1975) and Butler and Cobbold (1974). Planar catheter-tip oxygen sensors based on the Clark cell are described by Honig (1980), Engels and Kuypers (1983), Miyahara *et al* (1983), Sansen *et al* (1985), Koudelka and Grisel (1985), Kuypers (1985), Prohaska *et al* (1986) and Suzuki *et al* (1988). The sensor described in Engels and Kuypers (1983) has now evolved into an 'in vivo' device in a pre-commercial phase being made ready for the market by Dräger Medical Electronics of Best in the Netherlands. More details on this sensor are given in section 5.4.

All the planar realizations of the Clark cell cited above are two-electrode configurations where reference electrode and auxiliary electrode are combined into a single Ag/AgCl anode, as originally presented by Clark in 1956. Hence, the reference electrode is polarized and the potential may fluctuate if the area of the electrode is too small. Also, as seen in reaction (2.40), this Ag/AgCl electrode is consumed during the reduction of oxygen,

So, the operational lifetime of the sensor is restricted by the total amount of Ag present in the sensor. In planar devices in particular, this amount is very limited. Only a relatively thin layer ($< 3\ \mu m$) of Ag can be used with planar technology. The area of the anode is also limited, as the sensor has to fit into the catheter. These limitations result in a restricted lifetime of the sensor.

In Lambrechts (1989a) a new original solution is proposed for this problem. If a three-electrode system is used instead of a two-electrode system, a theoretical infinite lifetime is obtained. By proper choice of the auxiliary electrode metal, no material is consumed. In a study by the authors (Lambrechts 1989a), it was experimentally verified that a Pt auxiliary electrode suits this objective. After 24 hours continuous operation of a CMOS-compatible oxygen sensor, no damage of the thin-film Pt auxiliary electrode is seen. The exact reaction at the Pt auxiliary electrode is not known to us. It is supposed that the opposite reaction to (2.39), or a similar oxygen generation reaction, takes place. As an auxiliary electrode, an external stainless steel electrode, such as the housing of the catheter tip sensor or an injection needle, can also be used (Kuypers 1988).

Therefore, the three-electrode system has several important advantages. Because no current flows through the reference electrode, the potential is well defined. The reference electrode can also be small, as there is no risk of electrode polarization. If a Pt auxiliary electrode is used, the sensor is not exposed to material consumption and has a theoretically infinite lifetime. However, the three-electrode configuration has one important disadvantage. If the reference electrode fails or the connection to the reference electrode breaks, the potentiostat has no feedback signal and a high potential is applied to the auxiliary electrode. With some commercial potentiostats (EG&G PARC 1982) this potential can be as high as 100 V, delivering a maximum current of 1 A. For 'in vivo' applications such a potentiostat can be lethal, so special attention has to be paid to the design of a potentiostat for 'in vivo' applications. The maximum current has to be below $5\ \mu A$; above this current electrical stimulation of the patient occurs.

Although the proposed solution seems trivial, no other planar microelectronic-based three-electrode oxygen sensor has been reported in the literature up to now. One author reports on a three-electrode thick-film oxygen sensor with Au working and auxiliary electrodes (Karagounis *et al* 1986). The only reason mentioned in this paper for the use of a three-electrode sensor is the desire to have a reference electrode that can be isolated from participation in the cell reaction; the advantages of three-electrode systems to lifetime are not mentioned. The choice of a Au auxiliary electrode is also totally inadequate. In solutions containing chloride ions, such as blood or PBS, a Au electrode will chloridate and will also be consumed during the measurement.

It can be concluded that with the introduction of a third electrode into the Clark cell, an improved device with an extended lifetime can be realized. This extension is not necessary for classical wire and tube electrodes, as the amount of Ag is sufficient to guarantee a long lifetime. For planar miniature devices for 'in vivo' use, where the amount of Ag on the anode is limited, the third electrode is a necessity in order to obtain a satisfactory lifetime.

2.3.2.3 Glucose and biosensors. The Clark cell was also the source of inspiration for several other types of sensors based on the electrochemical reduction of oxygen. In combination with a glucose oxidase membrane, a glucose sensor can be made (Updike and Hicks 1967). The commercially successful Beckman glucose analyser is also based on this principle.

Under normal conditions, it is difficult to oxidize or reduce glucose electrochemically. It is, however, possible with a succesful technique known as pulsed amperometric detection (PAD) to condition the electrode surface in such a way that a stable and active electrode surface is created for the oxidation of, for example, carbohydrates such as glucose. A disadvantage of this technique is that carbohydrates can only be detected in high pH solutions (above pH 11) (Johnson 1986, Olechno *et al* 1987, Hardy and Townsend 1988, Rocklin *et al* 1990). This technique has been developed commercially by Dionex, Sunnyvale, USA in their special flow cell for pulsed amperometric detection in chromatography applications; separation of the different carbohydrates is accomplished in special chromatography columns (Dionex 1989). For biochemists this technique combines the possibility to differentiate and measure with high accuracy a very large variety of carbohydrates. If enzymatic biosensors were used for the same application, one would need a different enzyme for each selected carbohydrate. For medical applications, and especially for 'in vivo' techniques, this technique is not so suitable due to the high pH environment that is required and due to the lack of selectivity of the pulsed amperometric detection without chromatograpic separation.

There are also several research groups which have investigated the electrocatalytic reduction and oxidation of glucose at Pt black electrodes. It is seen that several peaks in the cyclic voltammogram of a Pt black electrode appear if glucose is added to a Krebs–Ringer buffer solution. These peaks are situated in the hydrogen region of the voltammogram. As the adsorption peaks in this region of the Pt electrode itself are not stable and depend on the electrode history, this method is very sensitive to electrode poisoning and electrode history. This becomes clear if the cyclic voltammograms in the different papers of the same author are compared (Yao *et al* 1982, 1983, 1984, 1986). Extensive variations are observed in the shape of the voltammograms shown in these papers. This proves that this method, although it works in principle, is limited in practical applications. Despite this important drawback, the electrocatalytic method has become more and more popular in the world of glucose sensors (Lewandowski *et al* 1986, 1987, Chan *et al* 1987, Sarangapani *et al* 1987).

The enzyme-based method for glucose sensing is, for medical applications, the most interesting approach. This method is based on the enzymatic conversion of glucose to gluconic acid and hydrogen peroxide in the presence of oxygen. This reaction is catalysed by the enzyme glucose oxidase (GOD or GOX):

$$\text{glucose} + O_2 \rightarrow \text{gluconic acid} + H_2O_2 \qquad (2.43)$$

$$\downarrow \qquad\qquad \downarrow \qquad\qquad \downarrow$$

$$\text{oxygen} \qquad \text{pH} \qquad H_2O_2$$

$$\text{sensor} \qquad \text{electrode} \quad \text{detector.}$$

As can be seen in the previous reaction, an increase in the glucose concentration will result in a decrease in the oxygen concentration, a decrease of the pH value and an increase in the H_2O_2 concentration. All these parameters can be used for the development of a glucose sensor. A suitable detector, covered with a glucose oxidase membrane, will respond to variations in the glucose concentration. The following detectors can be used.

1. Oxygen sensor

As oxygen is consumed during the enzymatic reaction, the decrease in the oxygen concentration in a glucose oxidase membrane is a linear function of the glucose concentration. The oxygen concentration can be measured with a Clark cell. With this method, the oxygen concentration in the sample also has to be measured, as this can change during the measurement. A differential set-up is needed (this technique is used by several research groups and is described, for example, in Gough *et al* (1982)), which has the disadvantage of therefore necessitating a more complicated device.

2. pH electrode

As gluconic acid is generated during the enzymatic reaction, the pH value will decrease as a function of the glucose concentration. The pH in the glucose oxidase membrane can be measured with a conventional pH electrode, or with an ISFET (see section 2.5). Since the pH value of the solution can change during the measurement, a differential set-up is also necessary in this case. The major disadvantage of this method however is the dependence of the response of the glucose sensor to changes in the buffer capacity of the solution (see also section 2.5).

3. H_2O_2 detector

As H_2O_2 is generated during the enzymatic reaction, the H_2O_2 concentration will increase linearly as a function of the glucose concentration. Normally, H_2O_2 is not present in samples to be analysed, so a single H_2O_2 detector in the glucose oxidase membrane is sufficient for the determination of the glucose concentration in the solution: no differential set-up is needed. A Pt electrode biased at 0.7 V versus a Ag/AgCl electrode can be used as H_2O_2 detector. This method, however, suffers from interferences of electro-oxidizable compounds in the solution, such as ascorbic acid, uric acid and Fe(II).

The disadvantage of the glucose oxidase-based glucose sensors is the dependence of the enzymatic reaction on the oxygen concentration. At high glucose concentration or low oxygen concentration, all oxygen in the GOD membrane is consumed. In this case, changes in the glucose concentration will not contribute to variations in, for example, the H_2O_2 concentration.

This is evidenced by a thorough examination of reaction (2.42). Glucose oxidase is a redox enzyme with a FAD centre (FAD = flavine adenine dinucleotide). Glucose reduces this FAD centre to $FADH_2$, the reduced form of FAD. During this reaction glucose is oxidized to gluconolactone:

$$\beta - D - glucose + GOD - FAD \rightarrow \delta - D - gluconolactone$$
$$+ GOD - FADH_2. \tag{2.44}$$

The oxygen in the solution oxidizes the $FADH_2$ centre back to FAD and H_2O_2 is formed:

$$O_2 + GOD - FADH_2 \rightarrow H_2O_2 + GOD - FAD. \tag{2.45}$$

This reaction regenerates the enzyme GOD back to its original state. Gluconolactone further reacts to gluconic acid:

$$\delta - D - gluconolactone + H_2O \rightarrow gluconic\ acid. \tag{2.46}$$

The H_2O_2 can be oxidized at a positively biased Pt electrode according to the following reaction:

$$H_2O_2 \rightarrow O_2 + 2H^+ + 2e^-. \tag{2.47}$$

During this reaction only a small part of the oxygen consumed in reaction (2.45) is regained, as not all the H_2O_2 formed in reaction (2.45) is oxidized at the electrode. A schematic view of the reactions taking place in the enzyme membrane is shown in figure 2.27. The overall process is influenced by the kinetics of the electrode reaction, by diffusion of the elements involved towards the electrode and by stoichiometric limitations. Generally a mixed-control process occurs in a glucose sensor.

1. Kinetic control
In the kinetically controlled process, the enzymatic reaction is not fast enough to oxidize all the glucose molecules which diffuse towards the electrode. The overall rate is determined by the kinetics of the enzyme reaction

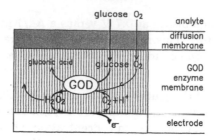

Figure 2.27   A schematic view of the reactions taking place in the glucose oxidase enzyme membrane.

and mass-transfer parameters do not influence the response. Kinetic control takes place at low enzyme concentrations and with permeable membranes. In practice, kinetic control will be avoided. The current is then proportional to the enzyme concentration and enzyme activity. Since the activity of glucose oxidase decreases as a function of time, the response of the sensor will also decrease. The response of the sensor is only linear with the glucose concentration in a limited concentration range; namely, as long as the glucose concentration is lower than the value of the Michaelis constant. A kinetically controlled process is described by the model of Michaelis–Menten (Stryer 1983).

2. Diffusion control

If the process is diffusion controlled, then all the glucose that diffuses towards the electrode is oxidized, since the kinetics of the reaction are fast. Diffusion control takes place at high enzyme concentrations and with a membrane of low permeability. This system is independent of the enzyme concentration and a linear response is obtained over a broad glucose concentration range. The major advantage of a diffusion-controlled system is that the response of the sensor is independent of the enzyme concentration and remains stable even if the enzyme activity decays. The sensor remains stable until the enzyme activity reaches a level where the process becomes kinetically controlled.

3. Stoichiometric control

The reaction can also be limited by stoichiometric restrictions. If the process is limited by oxygen (i.e. all oxygen is consumed during reaction (2.45)), the glucose sensor will become insensitive to glucose and will only respond to changes in the oxygen concentration. The oxygen deficiency results in saturation of the sensor for high glucose levels and limitation of the linear region. This is especially a problem for 'in vivo' sensors; the sample can not be diluted and the oxygen concentration is low (0.2 mM in arterial blood and even lower in tissue). A reliable glucose sensor must respond linearly up to 400 mg dl^{-1} glucose (22 mM) under such low oxygen concentrations.

Different solutions to this oxygen deficiency problem have been proposed in the literature.

1. A diffusion-restricted membrane can be placed on the top of the enzyme membrane. This membrane will slow down the diffusion of glucose towards the electrode. In this way, there is always enough oxygen in the enzyme membrane to allow reaction (2.45). Different materials can be used as a diffusion membrane. The best known are cellulose acetate, porous polycarbonate, cuprophan, polyurethane and Nafion. The main issue in the fabrication of these diffusion membranes is the reproducibility of the permeability and the thickness of the membrane, as these parameters define the sensitivity of the glucose sensor.

2. Another solution to this problem is to generate molecular oxygen in situ by the controlled electrolysis of water. A fourth electrode can be used for this purpose. This principle has been employed by Enfors (1981a,b) for the realization of an oxygen-insensitive probe for fermentation control. The main issue with this method is the exact control of the oxygen generation process; if too much oxygen is generated, oxygen gas bubbles will develop under the membrane which may result in destruction of the glucose sensor.

3. A third approach that is gaining interest is based on mediators. These chemical compounds form an alternative for oxygen in reaction (2.45). The best known mediator is ferrocene and its derivatives. If ferrocene is immobilized together with glucose oxidase in the enzyme membrane, the overall reaction (2.43) is replaced by

$$\text{glucose} + 2\,\text{ferricinium}^+ + H_2O \rightarrow \text{gluconicacid} + 2\,\text{ferrocene}. \quad (2.48)$$

At a positively biased electrode (160 mV versus SCE) the reduced couple is then reoxidized:

$$2\,\text{ferrocene} \rightarrow 2\,\text{ferricinium}^+ + 2e^-. \quad (2.49)$$

Glucose sensors based on this principle exhibit a linear range up to 30 mM, even with zero oxygen concentration solutions. A disadvantage for 'in vivo' use of mediators is the toxicity of these products. A commercial screen-printed glucose sensor for 'in vitro' glucose determinations based on this principle is now marketed under the name 'ExacTech' by Medisense. This commercial device originates from the research on mediators at Cranfield Biotechnology Centre (Turner 1988, Matthews *et al* 1987) and it is the only planar glucose sensor on the market at this moment. Combining a mediator as well as the enzyme into a carbon-paste electrode also results in working biosensors (Amine *et al* 1991). For more details concerning mediators, the reader is referred to the extensive literature on this topic (Janata and Bezegh 1988, Janata 1990, 1992).

4. Direct electron communication between the enzyme and the electrode can also solve the problem of oxygen deficiency. This method is generating

much interest (Janata 1990), as no (probably toxic) mediator is necessary. Two different techniques can be distinguished: the enzyme itself can be modified (Degani and Heller 1987), or the electrode surface can be modified in such a way that the oxidase enzyme can be reoxidized directly at the surface. An interesting development is described in Koopal *et al* (1991) where polypyrole, a conducting polymer, is polymerized in the holes of filter membranes. On this complex polypyrole surface, glucose oxidase is adsorbed. With this structure direct electron transfer between the enzyme and the polypyrole electrode is also demonstrated.

Based on this discussion it can be concluded that a glucose sensor has to be designed so that the device is working under diffusion control. In order to solve the oxygen deficiency problem, the best way to go seems to be the application of a diffusion membrane. Mediators and direct electron communication techniques are also very promising.

The first glucose sensors were rather clumsy and not suited for 'in vivo' use. They consisted of a Pt working electrode surrounded by a large Ag ring reference electrode in a two-electrode configuration. The membrane and the inner electrolyte were held in place with an 'O'-ring seal and a screw cap; the membrane itself was disposable and had to be changed daily. The electrode system was cleaned when the membrane was refurbished. This type of sensor is still in use in commercial glucose analysers from Yellow Springs Instrument Company, Ohio or in the Beckmann glucose analyser.

For 'in vivo' applications, several needle-type electrodes and catheter-tip electrodes have been developed by research groups all over the world. Well known needle-type sensors are described in, for example, Shichiri *et al* (1982, 1983), Churchouse *et al* (1986), Garcia *et al* (1983), Clark and Duggan (1982), Thévenot (1988). In particular, the needle electrode from Shichiri has found many adherents. This glucose sensor, shown in figure 2.28, consists of a Pt-ball working electrode and an annular Ag electrode sealed in glass. The enzyme glucose oxidase is immobilized in a bovine serum albumin matrix. To increase the linear region, a polyurethane membrane is used. A similar device, developed by Colin *et al* (Garcia *et al* 1983) at the ULB, Belgium is shown in figure 2.29. This three-electrode glucose sensor consists of a Pt ball working electrode, also surrounded by an annular Ag electrode. The enzyme is likewise immobilized in a bovine serum albumin matrix. To increase the linear region and exclude large molecules such as proteins, a porous polycarbonate diffusion membrane is used. The two electrodes, with the membrane on top of it, are placed in an injection needle. This injection needle also serves as an auxiliary electrode. The problem with these needle-type electrodes is the costly manual production technique. The isolation of the thin Pt wires (200 μm diameter) in the small Ag tubes (400 μm internal diameter) is a particular bottle-neck. Also, the reproducibility of the membrane thickness and permeability is difficult

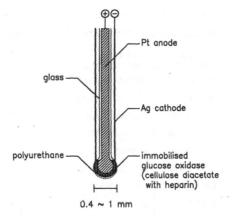

Figure 2.28   Needle-type glucose sensor as described in Shichiri *et al* (1982).

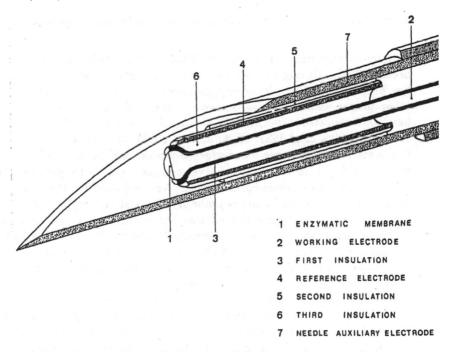

1 ENZYMATIC   MEMBRANE

2 WORKING ELECTRODE

3 FIRST   INSULATION

4 REFERENCE ELECTRODE

5 SECOND   INSULATION

6 THIRD   INSULATION

7 NEEDLE AUXILIARY ELECTRODE

Figure 2.29   Needle-type glucose sensor as described in Garcia *et al* (1983).

to control with manual production techniques. None of these needle-type electrodes are commercially available.

The realization of planar voltammetric glucose sensors with microelectronic production techniques was, for the authors, the start of their work

Table 2.3   A survey of some typical enzyme biosensors.

determinant	substrate	enzyme	detector
saccharides	glucose	glucose oxidase	O_2, H_2O_2 electrode
alcohols	ethanol	ethanol oxidase	O_2 electrode
aminoacids	glutamate	glutamate dehydrogenase	NH_4^+ electrode
acids	acetic acid	alcohol oxidase	O_2 electrode
	uric acid	uricase	H_2O_2 electrode
lipids	cholesterol	cholesterol oxidase	H_2O_2 electrode
antibiotics	penicillin	penicillinase	pH electrode
other substrates	urea	urease	NH_4^+, pH electrode
	creatinine	creatininase	NH_4^+, pH electrode
	inorganic phophorus	alkaline phosphatase	H_2O_2 electrode

on microelectrochemical biosensors (Sansen *et al* 1985, Sansen and Lam-brechts 1985). During the development of this glucose sensor, the interest has been enlarged to planar oxygen sensors, ion-selective electrodes and other planar electrochemical sensors. Meanwhile other research groups have also reported experimental planar glucose sensors (Miyahara *et al* 1983, Koudelka *et al* 1987, Kimura *et al* 1988). It is believed that within five years microelectronic-based glucose sensors will be available for short term (2–3 days) 'in vivo' glucose-level control for diabetic patients. This sensor will be an enzyme-based device. As a solution for the oxygen deficiency, the use of mediators is probably the best choice, provided that a biocompatible, FDA-approved mediator can be found. As an alternative, diffusion membranes have to be developed. The main problem with these diffusion membranes, and also with enzyme membranes, is finding a way to prepare them with microelectronic-compatible techniques. Some new methods will be discussed in this work.

A glucose sensor is a typical example of a biosensor (i.e. a sensor made with a biological selective substance). A complete set of sensors can be derived by changing the active enzyme. A selected overview of the different possibilities is given in table 2.3. If the appropriate enzyme is available, every clinically significant component can be measured in principle with a modified glucose sensor. Only a change of the chemically active membrane is necessary. Because of the high selectivity and diversity of an enzyme reaction (e.g. glucose oxidase will only react with β-D-glucose, not with α-D-glucose or other sugars), extremely selective devices can be made for a number of substrates if the corresponding enzyme can be found in nature or can be synthesized artificially.

2.4 CONDUCTOMETRY

2.4.1 Theory

Electrolytic conductance is a non-faradaic process that can give useful chemical information. The origin of electrolytic conductance is the transport of anions to the anode and cations to the cathode. In order to complete the current path, electrons are transferred at the electrode surface to and from the ions. The conductance of an electrolyte is measured in a conductance cell consisting of two identical non-polarizable electrodes. To prevent polarization, an AC potential is applied to these electrodes and the AC current is measured.

Conductance (L) is defined as the current (A) divided by the potential (V). The unit of conductance is Ω^{-1}, or Siemens. The reciprocal of conductance is resistance (Ω).

Conductivity (κ), or specific conductance, is defined as the current density ($A\,m^{-2}$) divided by the electrical field strength ($V\,m^{-1}$). The unit of conductivity is ($\Omega^{-1}\,m^{-1}$). The reciprocal of conductivity is resistivity ($\Omega\,m$). The conductivity can be thought of as the conductance of a cube of the electrolyte solution with unit dimensions.

As the conductivity is defined as the conductance of a cell formed by two electrodes with unity area at unity distance, a cell constant is introduced for the normalization of different conductance cells. The cell constant K is the product of the known conductivity κ with the measured resistance R of the solution:

$$K = \kappa L^{-1} = \kappa R. \tag{2.50}$$

The cell constant can be seen as a calibration factor for a certain electrode configuration. For simple electrode configurations, this constant can be calculated theoretically.

In figure 2.30 a conductance cell and its equivalent electric circuit are shown. The goal of conductometry is the determination of the solution resistance R_{sol}. Resistance R_l represents the lead wire resistance which can be neglected in most applications. The electrode impedance consists of two elements: the double-layer capacitance C_d and the faradaic impedance Z. Capacitance C_p is the interelectrode capacitance.

By applying an AC potential, an AC current will flow through C_{d1}, R_{sol} and C_{d2} and in parallel through C_p. If the double-layer capacitance is sufficiently large, no potential will build up across the corresponding faradaic impedance so high such that faradaic currents can flow. So, if C_p can be kept small, only R_{sol} contributes to the measured conductance. In contrast with potentiometry and voltammetry, conductometry monitors processes in the bulk of the solution. Any influence of electrode processes has to be avoided.

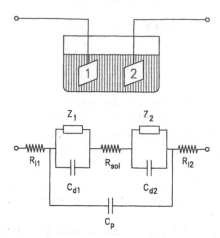

Figure 2.30   A conductance cell and the corresponding electric circuit.

The conductivity of an electrolyte solution is equal to the sum of the conductivities of each type of ion. The term equivalent conductance (Λ) is introduced to allow comparison of the conductivity of different electrolytes; it gives a relation between the conductivity and the concentration of a specific ion. It is normalized to the charge and the concentration of the ion. The equivalent conductivity Λ is defined as the conductivity of the solution multiplied by the solution volume that contains one gram-equivalent of the electrolyte.

The equivalent conductivity Λ ($\Omega\ cm^2$) therefore becomes

$$\Lambda = \frac{10^{-3}}{zC}\kappa. \qquad (2.51)$$

In practice, one measures the solution resistance R and calculates the conductance as $\Lambda = 1/R$. Taking into account formula (2.50), one can calculate the conductivity as $\kappa = K\Lambda$ and finally one obtains Λ from formula (2.50). For some applications a four-point electrode configuration is preferred to a two-point electrode. More details can be found in section 3.3.

2.4.2   Applications

Conductometry is extensively used for the determination of water purity. Ultra-pure water, as used in the semiconductor industry, can have a resistivity up to 18 Ω cm (distilled water has a resistivity of 1 Ω cm), so conductometric cells are placed almost everywhere in the water purification system in order to evaluate the process.

Conductometry can also be used in tritrations and as an electrochemical detector after chromatographic separations. In the scope of this book it is also important to notice that conductometric biosensors can be realized. More details can be found in section 5.3.5

2.5   ION-SENSITIVE FIELD EFFECT TRANSISTORS

Ion-sensitive field effect transistors (ISFET) are a special type of MOSFET transistors, where the gate oxide of the device is in direct contact with an electrolyte. This device is also known as CHEMFET (chemically sensitive field effect transistor). The first report of an ISFET is a short communication by Bergveld in 1970 (Bergveld 1970), followed by a more extensive paper in 1972 (Bergveld 1972). He developed the device initially for measurement of pH and sodium ion activity, as well as for recording physiological transient action potentials. A similar device was reported in 1971 by Matsuo (Matsuo and Wise 1974). The exact theoretical principles of the ISFET were not known at that time. Several papers describing the theoretical background of ISFETs were published a few years later (Janata and Huber 1980, Kelly 1977, Buck and Hackleman 1977). At that moment an intensive debate was taking place on the need for a reference electrode. In, for example, Bergveld (1970, 1972) and Zemel (1975) it is claimed that ISFETs can operate without a reference electrode. This claim has been contested by several authors (Moss *et al* 1975, Kelly 1977 and Buck and Hackleman 1977). It is now generally accepted that a reference electrode is needed for the stable operation of an ISFETs. It is believed that the initial ISFETs without a reference electrode worked due to small leakage currents through the gate oxide or the package, resulting in a second reference interface with the electrolyte. The potential over this interface, however, is undefined and unstable resulting in a varying response.

The ISFET has been investigated thoroughly, resulting in a large pile of publications and a great variety of devices such as enzyme FETs (ENFET), immunochemically sensitive FETs (IMFET), BIOFETs and reference FETs (REFET). Up to now, papers on the fabrication of the ISFET and on theoretical backgrounds are published regularly. For further details the reader is referred to the recent proceedings of the international conferences on solidstate sensors, such as Fukuoka'83, Philadelphia'85, Bordeaux'86, Tokyo'87 and Montreux'89. Despite all these efforts, no commercial product has appeared on the market yet. This is mostly related to drift problems of the device. In this section the basic principles of the ISFET will be detailed and a comparison between ISFETs and voltammetric sensors will be made.

2.5.1   Theory

In figure 2.31 a cross section of a typical ISFET is shown. It consists of a normal MOS transistor where the gate material (polysilicon or Al) is

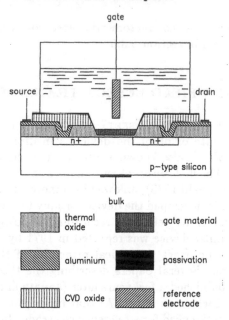

gate

source

drain

n+   n+

p—type silicon

bulk

	thermal oxide		gate material
	aluminium		passivation
	CVD oxide		reference electrode

Figure 2.31   A cross section of a typical ISFET.

removed and the gate oxide is exposed to the liquid. Usually the gate oxide layer is covered with a Si_3N_4 or Al_2O_3 layer to improve the characteristics of the ISFET such as stability, blood compatibility and Nernstian behaviour.

For operation in the saturation region, the following equation for the drain–source current I_{DS} can be derived (Janata and Huber 1980):

$$I_{DS} = \frac{\mu n C_{ox} W}{L} \left(V_G - V_T^* + \frac{RT}{nF} \ln(a_i) - E_{ref} \right)^2. \tag{2.52}$$

For the linear-region operation, one finds

$$I_{DS} = \frac{\mu n C_{ox} W V_D}{L} \left(V_G - V_T^* + \frac{RT}{nF} \ln(a_i) - E_{ref} - \frac{V_D}{2} \right) \tag{2.53}$$

where μ_n is the mobility of the electrons in the channel (cm^2 V^{-1} s^{-1}). C_{ox} is the capacitance per unit area of the gate insulator (pF cm^{-2}), W is the channel width (μm), L is the channel length (μm), V_G is the gate voltage (V), V_T^* is the threshold voltage of the ISFET (V), a_i is the activity of the ions where the ISFET is sensitive to them (M), E_{ref} is the potential of the reference electrode (V) and V_D is the drain voltage (V).

These equations are similar to the equivalent equations of a normal MOS transistor. One can distinguish two extra terms due to the potential of the reference electrode and due to the potential over the gate oxide/electrolyte interface. This potential depends on the ion activity in a similar way as in the Nernst equation. The response of an ISFET is therefore similar to potentiometric devices, i.e. the potential is a logarithmic function of the ion activity. The origin of the potential at the gate oxide/electrolyte interface has been studied extensively by several authors. A further discussion of this topic goes beyond the scope of this book. For more details the reader is referred to the literature.

2.5.2   Applications

ISFETs are in most cases used and developed for pH measurements. 'In vivo' pH measurement of blood is the primary objective. As a gate insulator material, SiO_2, Si_3N_4 or Al_2O_3, or a combination of these materials, is used. SiO_2 has the disadvantages that the oxide hydrates and that microcracks in the oxide layer result in large leakage currents through the gate. This leakage current results in drift of the device. The selectivity of this material is also not so good. Si_3N_4 is used for its excellent characteristics concerning leakage current and reproducibility. Si_3N_4 ISFETs also have a wider range of pH sensitivity. Al_2O_3 ISFETs are more suited for 'in vivo' measurements.

The most important problem of the ISFET is the drift of the device. Initially, the drift problem was related to packaging problems. All leakage currents from the gate or the source, drain and bulk regions have to be avoided. Several solutions for the packaging problems have been proposed, beside the classical epoxy and silicon rubber techniques: TAB and dry-film photoresist techniques (Ho *et al* 1983), silicon-on-sapphire (SOS) devices (Kimura *et al* 1985), glass encapsulation (van den Vlekkert *et al* 1987) and backside contacts (van den Vlekkert 1988). The backside contacts look especially promising for the definitive solution of the packaging problem. One disadvantage of this method is the relatively large size of the contacts (480 μm by 480 μm), excluding the use of this technique for the connection of more complicated integrated sensors.

After solving the packaging problem some drift of the device still remains. The fact that this drift and its specific characteristics are reported by practically all authors on ISFETs indicates that the drift of the ISFET is a fundamental problem of the device. Several studies on the drift problem have been published and, as solution, pulse techniques and a drift prediction based on measured characteristics of the device have been proposed [van den Vlekkert 1986, Ligtenberg 1987]. An effort to develop commercially an ISFET by the Dutch company Sentron using drift correction built into an EPROM in the connector of the device is still going on.

From these remarks on pH ISFETs it can be concluded that at the moment these devices can only be used for applications where the high precision necessary for blood pH analysis is not a pre-requisite. Such applications are the measurement of pH in the stomach or the pH of dental plaque. The clinically significant range varies for these applications over several units of pH, instead of a few tenths for blood analysis.

Also, ISFETs for other ions, such as Ca^{2+}, K^+ and Na^+, have been made (Janata and Huber 1980, Sibbald *et al* 1985, Tsukada and Goffe 1987). These devices are based on ISFETs, as described above, covered with an ion-selective membrane such as a PVC membrane with valinomicin for the measurement of K^+ ions. They detect changes in the charge of the membrane, or the potential over the membrane. From the theoretical as well as from the practical side, these devices have several important drawbacks. Several research groups working on ISFETs for Ca^{2+}, K^+, Na^+ and similar ions have discontinued their research due to the improper coupling mechanism resulting in a substantially reduced overall selectivity or a total loss of the response (Janata 1986) .

Enzyme FETs (or ENFETs) have also been developed (Kimura *et al* 1985, Miyahara and Moriizumi 1985, Näbauer *et al* 1986). A typical application is a miniature 'in vivo' glucose or urea sensor. These sensors are based on a change in pH in an enzyme membrane by a chemical reaction, such as that described in formula (2.42). An increase in the glucose concentration will in this case result in an increase in the gluconic acid concentration and hence in a decrease of the pH in the enzyme membrane. As a response to a varying glucose concentration, a typical S-shaped curve is obtained. In the first region of the curve the pH change in the membrane is buffered by the buffer capacity of the electrolyte (PBS or blood). If this buffer capacity is exceeded, the pH in the membrane will decrease and a change in the potential of the ENFET is measured; in this region the potential is a linear function of the glucose concentration. At higher glucose concentration, saturation is observed because of, for example, oxygen deficiency in the membrane.

These ENFETs have some important drawbacks. The shape of the response curve is extremely sensitive to changes in the buffer capacity or the pH of the measurement solution. Although this phenomenon has been described experimentally and theoretically (Eddowes 1987), no good answer to these problems has been found (Janata and Bezegh 1988). A voltammetric sensor, detecting O_2 or H_2O_2, is therefore a preferable alternative.

Immunochemically sensitive FETs (IMFET) are also described in the literature (Janata and Huber 1980, Janata 1986). These sensors are based on the covalent coupling of, for example, a specific antibody to the gate area of an ISFET. When immersed in a solution containing the corresponding antigen, antibody and antigen will couple, resulting in a change of the charge above the gate of the ISFET. A fundamental problem with these devices is the fact that this change in charge is taking place outside the outer Helmholtz plane (OHP) (see section 2.1.2) due to the large size of the immunological molecules involved, so all change in charge is screened by the non-specifically adsorbed ions at the OHP. Hence, in principal these devices cannot work as anticipated. The exact mechanism why these IMFETs should, or should not, work is not yet understood at all.

It can be concluded that ISFETs are well suited for pH measurements where the drift problem does not interfere with the application. At first glance, such applications are the measurement of pH in the stomach or the pH of dental plaque. Also, several industrial processes do not require an extremely precise pH measurement. ISFETs can be adopted for these applications. The derived devices, such as ENFETs and IMFETs, are, however, unreliable where classical electrochemical detection techniques are preferable alternatives.

An appreciable benefit of the ISFET is certainly the international attention that this device has drawn to solid-state sensors in the academic world, as well as in the industrial and commercial world. Without the ISFET, the knowledge of solid-state/electrolyte interfaces, integrated chemical sensors and sensor packaging technology would still be in its infancy. The question arises why all this research has not resulted in a commercial device. Often a comparison is made with the charge-coupled devices (CCD); CCDs were also invented in between the the sixties and the seventies (Boyle and Smith 1970, Tompsett *et al* 1971). Nowadays, CCDs can be found in several consumer products, such as video cameras and autofocus systems in reflex cameras. For special applications CCDs, are available with more than 1 000 000 pixels. Also, new theories and special packaging techniques have been derived during the development of reliable CCDs.

The number of phases involved, however, is the main difference with ISFETs. They deal with two phases: the solid-state phase and the liquid phase. For the development of ISFETs, electronics and electrochemistry have to be combined. This makes experimental and theoretical work more demanding. CCDs also have direct military and space applications, whereas ISFETs are oriented more towards biomedical applications. Therefore, IS-FETs did not receive the extra impulse found in military and space projects.

It can be concluded that the ISFET is not dead, although some people like to refer to it as the WASFET. The drift problem prevents the use of this device at the moment for 'in vivo' blood pH measurement, but it is believed that soon new applications will be found where the ISFET can be used on a commercial basis.

Possible directions are indicated in a recent article of the inventor by the ISFET (Bergveld 1991). In this paper, it is stated that the ISFET should be used in situations were its fast response time and small dimensions are fully exploited. Bergveld suggests concentrating on dynamic measurements instead of static measurements. Dynamic ISFET systems which have already been demonstrated are flow injection systems, ion-step immunofets and coulometric titration systems.

2.5.3 A comparison between ISFETs and microelectrochemical sensors

As this work details mainly planar voltammetric sensors, a comparison between ISFETs and voltammetric sensors is imperative. The ISFET is a completely new electrochemical device that is based on the intrinsic characteristics of the semiconducting material, silicon. Hence, an ISFET is an integrated chemical sensor. Microelectrochemical sensors are also usually made on a silicon substrate. The silicon does not, however, take part in the sensing process; it only serves as a smooth, flat surface for the deposition of the planar noble-metal electrodes. Silicon technology was chosen for the possibility of integrating interface electronics on the same chip as the sensor, resulting in a smart sensor. The only role of these interface electronics is to amplify or condition the weak signal of the sensor. This separation of the electronic part from the electrochemical part has the intrinsic advantage that part can be optimized without conflict of interests.

ISFETs are also restricted to potentiometric-like measurements. By integration or planarization of more classical electrochemical electrodes, the complete range of electroanalytical techniques is available; potentiometry, voltammetry, amperometry, conductometry,... .

ISFETs, due to the potentiometric-like characteristics, give a logarithmic response of the potential as a function of the concentration. This is an interesting type of response if a broad range of concentrations has to be measured. Voltammetric sensors give a linear response of the current as a function of the concentration. This is an interesting type of response if a limited range of concentrations has to be measured with a high accuracy, as is the case for 'in vivo' glucose sensing, for example.

A disadvantage of miniature electrochemical sensors is the small current and the high impedance of the signal. The outputs of an ISFET (the source and the drain) are low-impedance points. This implies that microelectrochemical sensors are sensitive to electromagnetic interferences. The signal line between the sensor and the interface electronics has thus to be kept short. ISFETs, on the other hand, are very sensitive to electrostatic discharges; an electrostatic discharge can damage the gate insulator.

For the utilization anticipated in this book, the most important disadvantage of the ISFET is that the application is mainly limited to pH sensors. Since the goal of this work is the realization of different types of microelectrochemical biosensors for various ions, it is clear that microelectrochemical sensors are the way to go.

2.6 CONCLUSION

In this chapter the basic principles of electroanalytical chemistry, applied to microelectrochemical sensors, are reviewed. Starting with a discussion on electrode processes, definitions and terminology are clarified. Since several electrochemical phenomena find their origin in the electrochemical double layer, special attention is paid to this matter. The Nernst equation is derived and applications of potentiometry are presented.

The most important part of this chapter deals with voltammetry. Voltammetry was defined as an electroanalytical technique where the current through the cell is recorded as a function of the applied potential. Current–potential relationships are measured at an electrode immersed in a solution containing electroactive species. The measured current is linearly proportional to the concentration of the electroactive species of interest. Different real-life experiments are described and related to theoretical principles. Two important applications of voltammetry are detailed, i.e. the Clark cell for dissolved-oxygen sensing and enzymatic glucose sensors.

Conductometry, a third electroanalytical technique, is described and applications are detailed.

This chapter is completed by a discussion on ISFETs (ion-sensitive field effect transistors). The advantages and disadvantages of microelectrochemical sensors and ISFETs are summarized. It is concluded that for the realization of miniature biosensors, microelectrochemical sensors are the way to go.

CHAPTER 3

MEASURING TECHNIQUES FOR SENSOR EVALUATION

Measuring techniques have two different aspects. Firstly, in order to evaluate the sensor, its response must be characterized. This evaluation is an especially intensive task during the development of a sensor. In this development phase, as many parameters as possible need to be characterized. Complex measuring techniques and expensive equipment are therefore necessary.

In the second phase, sensors have to be used by the end user: a physician, nurse or laboratory assistant, or even the patient himself. The measuring technique and equipment have to be inexpensive, user friendly and fool proof. In this end user-product, calibration procedures also have to be implemented in a transparent and ergonomic way.

As this book deals with the development of microelectrochemical sensors, most of the attention is directed towards evaluation systems for microelectrochemical sensors. However, the end-user application has to be kept in mind so that techniques developed during the evaluation phase can easily be transferred to the application phase.

In this chapter, three electroanalytical measuring techniques are described: potentiometry, voltammetry and conductometry.

3.1 POTENTIOMETRIC TECHNIQUES

In the previous chapter, potentiometry is defined as the measurement of the potential of non-polarized electrodes under conditions of zero current. Prior to the existence of modern electronic equipment, a potentiometer was used for the determination of the potential. The diagram of a potentiometer with a linear voltage divider is shown in figure 3.1. The system consists of a battery connected to the linear voltage divider. The resistance from point A to C is directly proportional to the length AC of the resistor. A galvanometer, G, is used as a zero-current detector. This potentiometer is

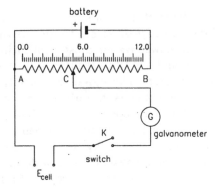

Figure 3.1 Diagram of a potentiometer with a linear voltage divider.

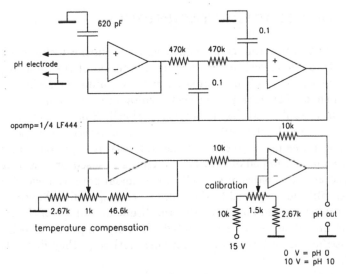

Source : databook National Semiconductor Corporation

Figure 3.2 Circuit diagram of an inexpensive pH meter.

calibrated using a standard cell such as the Weston standard cell. Firstly, the standard cell (with a standard potential E_s) is connected to the circuit and the distance AC_s where no current flows through the galvanometer is determined. In the second step of the measurement, the unknown cell is connected to the potentiometer and the distance AC_x where no current flows is determined again. The potential of the cell can be calculated as

$$E_x = E_s \frac{AC_x}{AC_s}. \tag{3.1}$$

A measurement of the electrode potential with a potentiometer is rather complex. Therefore, direct-reading electronic instruments for potential measurements are used nowadays for most applications. These electronic pH meters consist of a high (> 10 GΩ) input impedance amplifier and an analogue-to-digital convertor. The signal can usually be displayed in mV or in pH. With two potentiometers, the pH meter can be calibrated at two points for offset as well as for slope variations. Usually some temperature compensation can be applied to the measurement, either by setting the temperature manually on a dial or by using an electronic thermometer. These pH meters are nowadays available in a wide range of precision, price and automation. In figure 3.2 a circuit diagram of a low-cost pH meter is shown.

3.2   VOLTAMMETRIC TECHNIQUES

In chapter 2, voltammetry is defined as a measuring technique where the current through the cell is recorded as a function of the applied potential. According to the applied potential waveform, different techniques can be distinguished. For example, in cyclic voltammetry a triangular potential waveform is applied to the electrochemical cell. Square wave voltammetry is a good example of a pulse method. Amperometry is a special case of voltammetry, as in this technique the potential is kept constant as a function of time. AC impedance measurements are based on the measurement of the AC characteristics of the cell: a sinusoidal potential waveform with a small amplitude is applied to the cell, the corresponding sinusoidal current waveform is recorded and, from the amplitude and phase of the current signal, the impedance can be calculated. Usually, impedance techniques are not classified as voltammetric techniques, although they fit in the general definition cited above.

Each technique has advantages, disadvantages and some typical applications (EG&G PARC 1982); these are discussed in the next sections. In figure 3.3 an overview of the different techniques and the corresponding concentration range is shown. From this figure it can be concluded that voltammetry is appropriate for the determination of high, as well as extremely low, concentrations. However, the techniques used for the determination of extremely low concentrations require expensive equipment and difficult procedures. For the development of commercially viable sensors, elementary techniques are preferred. In this case, voltammetry can be used for the measurement of concentrations as low as 10^{-5} M.

For the evaluation of planar voltammetric sensors, two different systems are described in this book: an elementary manual system and a computer-controlled measurement set-up. An evaluation system has to be capable of applying the required potential waveform to the voltammetric sensor

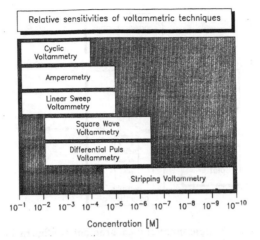

Figure 3.3   An overview of the different voltammetric techniques.

and measuring the current. From a voltammetric experiment, the current–potential curve (or voltammogram plot) is obtained.

3.2.1   Voltammetric measuring equipment: the potentiostat

The elementary manual system consists of a potentiostat, a function generator and an XY plotter (figure 3.4). The function generator generates the desired potential waveform and the XY plotter generates the voltammograms. The potentiostat itself is shown in figure 3.5. This potentiostat consists of four CA3140 operational amplifiers with MOS transistor input and a few resistors for the adjustment of the current range. Four current ranges are available: 100 μA, 10 μA, 1 μA and 100 nA full- scale. Although this potentiostat has excellent stability and good accuracy, the total evaluation system has several limitations. Only triangular, block or sine waves can be applied to the electrode. The reproducibility of the settings of the function generator is also insufficient. The total measuring procedure is labour intensive and does not allow post-processing of the measured data. Therefore, a computer controlled evaluation system is preferred for intensive use.

As a computer-controlled measurement set-up for the evaluation of voltammetric sensors, the authors have selected a PARC EG&G 273 potentiostat/galvanostat as the central piece of equipment. It can be interfaced to a computer with HP-IB (IEEE-488) or RS-232 interfaces. The potentiostat has an extensive command-set through which the system can be programmed. It has a high-compliance voltage (100 V), several filters to reduce noise, a high-speed or a high-stability mode and eight current ranges from 1 A to 100 nA full-scale. In the system, two 14 bit D/A convertors,

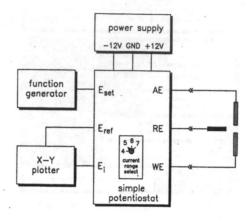

Figure 3.4   Simple manual measurement configuration for voltammetric sensors.

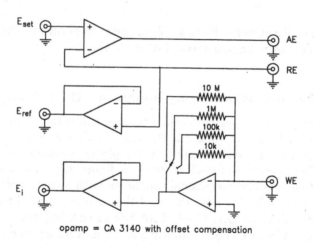

Figure 3.5   Circuit diagram of the simple potentiostat.

one 12 bit A/D convertor and 12k bytes of memory (to store data points) are incorporated.

The complete computer-controlled evaluation set-up is shown in figure 3.6. It consists of the following.

1. PARC EG&G 273 potentiostat/galvanostat

The PARC EG&G 273 is only used in the potentiostat mode. It is the heart of the complete evaluation system.

2. HP 9826 controller

This table-top computer has powerful capabilities for the control of HP-IB equipment. This computer controls all the digital equipment via its two HP-IB interfaces.

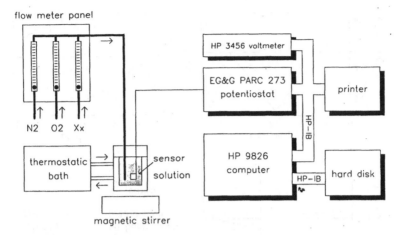

Figure 3.6   Configuration of the computer-controlled measurement set-up for evaluation of voltammetric sensors based on an EG&G PARC 273 potentiostat.

3. HP 82906A printer
This printer is used to obtain hard copies of the results.
4. HP 9133 hard disk, $3\frac{1}{2}$" floppy disk
All the different programs and program libraries are stored on the hard disk. The measured data points are stored on $3\frac{1}{2}$" in floppies.
5. HP 3456 voltmeter
This digital voltmeter is also controlled via the HB-IB bus. It is used for the synchronous measurement of temperature, humidity or other external parameters and for the evaluation of potentiometric and conductometric sensors.
6. Solution conditioning
The solution is contained in a special container for electrochemical experiments. The temperature of the electrolyte can be controlled between 20°C and 95°C using a thermostatic bath. The solution can also be stirred with a magnetic stirrer.
7. flow meter panel
For the evaluation of gas sensors and dissolved oxygen sensors, a flow-meter panel has been installed for mixing three gases. A third gas (e.g. diluted CH_4 or H_2) can be mixed with O_2 and N_2 in any proportion. This gas mixture can be bubbled through the solution or sent over the solution with a three-way valve in order to obtain a constant gas concentration in the solution.

A typical experiment is performed as follows: firstly, the experiment is programmed on the table-top computer. The computer translates the parameters of the desired experiment into commands for the EG&G PARC

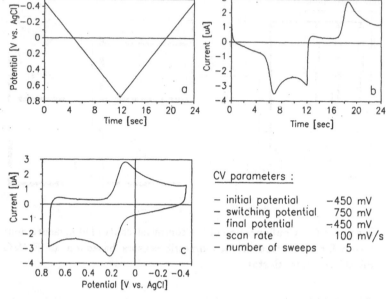

Figure 3.7   Potential (*a*), current (*b*) and voltammogram (*c*) of a typical cyclic voltammetry experiment.

273 potentiostat. Once the experiment is started, the computer is not strictly necessary as all experimental data is stored in the memory of the potentiostat. When the measurement is finished, the experimental data is transferred to the computer. On the computer various kinds of data post-processing can be implemented, such as plotting, filing and data comparison. For the system described in figure 3.6, a complete software package has been developed for different electrochemical techniques. These techniques are described in the next sections.

3.2.2   Cyclic voltammetry

In cyclic voltammetry (CV) the applied potential is changed linearly with time, starting from a potential where no electrode reaction occurs and moving to a potential where reduction or oxidation of the electroactive species involved occurs. After traversing the potential region where the electrode reactions take place, the scan direction is reversed and usually the electrode reactions of intermediates (i.e. products formed during the forward scan) are detected (Evans *et al* 1980). Cyclic voltammetry is an electrochemical technique that is used to study the kinetics of electrode reactions. It is a very popular technique for initial and mechanistic electrochemical studies of new systems. It is generally not used for analytical applications.

A triangular potential waveform is applied to the electrochemical cell (figure 3.7(a)). The scan rate of a cyclic voltammetry experiment can be varied typically over the range from 10 mV sec^{-1} to 100 V sec^{-1}. The peak-to-peak value of the applied potential wave will usually not exceed 2 V for aqueous electrolytes. The DC value of the wave depends on the electrolyte and the potential of the reference electrode. With a Ag/AgCl reference electrode in PBS, values around 0 V are used. A cyclic voltammetry experiment is completely defined by the scan rate, initial potential, switching potential, final potential and the total number of sweeps (one sweep is a complete period of the triangular wave). In most applications the initial potential equals the final potential.

The measured current at the working electrode (figure 3.7(b)) is the response of a cyclic voltammetry experiment. As discussed in section 2.3.1, a peak in the current is observed due to the combined effect of the decrease of the electrode surface concentration and the expansion of the diffusion layer with time. In the reverse scan a peak is also observed; this peak is due to the electrochemical reaction of intermediates or products of the reactions during the forward scan. In figure 3.7(c) the resulting voltammogram (current–potential curve) is shown. The important parameters of this cyclic voltammogram are: the cathodic and anodic peak potentials, the cathodic and anodic peak currents, the cathodic half-peak potential and the half-wave potential. The half-wave potential is usually situated within a few mV of the formal potential and provides valuable qualitative information on the electrochemical system involved. The potential peak separation also gives information on the reversibility of the electrode reaction. The difference equals 59 mV n^{-1} for a reversible system.

For sensor applications, cyclic voltammetry can be used for the initial evaluation of new electrode materials or new technologies for electrode fabrication. With cyclic voltammetry experiments, the residual current of an electrode can easily be determined. In addition, leakage currents due to defective packaging can be detected. It is also a powerful tool for the investigation of adsorption effects.

As described in section 5.2.1, cyclic voltammograms of sputtered Pt electrodes in H_2SO_4 have been recorded in order to compare the electrochemical characteristics of sputtered Pt with bulk Pt. Contamination by Ag was found on some Pt electrodes due to migration effects. From these voltammograms, and especially from the shape of the hydrogen region, the real active electrode area can be calculated (Woods 1976). By measuring the influence of the scan rate on the current response, the origin of the current can be determined (see also section 2.3.1).

3.2.3 Linear-sweep voltammetry

The analytical counterpart of cyclic voltammetry is linear-sweep voltammetry (LSV). In linear-sweep voltammetry experiments, a potential waveform

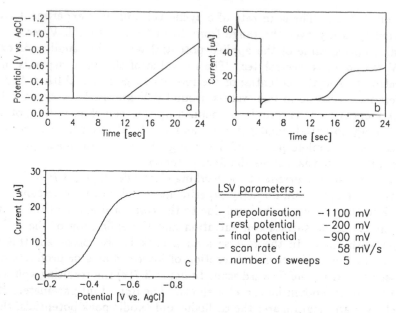

Figure 3.8 Potential (*a*), current (*b*) and voltammogram (*c*) of a typical linear-sweep voltammetry experiment.

consisting of pre-polarization pulses and linear potential sweeps is applied to the electrode system. A typical linear-sweep voltammetry experiment is shown in figure 3.8. The applied potential starts with a pre-polarization pulse to clean the electrode. This pulse is followed by a rest potential. During this rest period, the reaction products from the pre-polarization pulse surrounding the electrode can diffuse away, so that a fresh solution is found in the vicinity of the electrode. Measurement results are more reproducible in this way. The rest period is then followed by the real measurement, where a linear potential sweep is applied. The current during this potential sweep contains the analytical information (figure 3.8(*b*)). The linear-sweep voltammogram is then plotted as the current–potential curve during this linear potential sweep (figure 3.8(*c*)).

Linear-sweep voltammetry is an important technique for the evaluation of oxygen and glucose sensors. Pre-polarization techniques are essential for the conditioning of the electrode surface when Pt electrodes are used. Linear-sweep voltammetry has the advantage in contrast to amperometry, which is also a simple technique, that the complete current–potential information is recorded. Beside the concentration of the electroactive species involved, valuable information on the sensor condition can be extracted. From the shape of the voltammogram, problems with the background current, electrode poisoning and defective packaging can be traced.

Linear-sweep voltammetry is a universal technique that can be used in

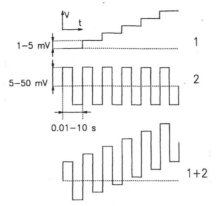

Figure 3.9 The applied potential waveform during square-wave voltammetry.

practically all analytical applications concerning microvoltammetric sensors.

3.2.4 Square-wave voltammetry

Square-wave voltammetry (SWV) is an example of a pulsed voltammetric technique. The foremost advantage of these pulse techniques is the capability to discriminate against capacitive currents. This results in methods more sensitive to faradaic currents than conventional techniques. Because of their differentiating characteristics, pulsed voltammetric techniques allow us to obtain nicely separated peaks for faradaic processes, instead of the S-shaped (sigmoidal) voltammograms obtained with normal voltammetric techniques. As a result, improved resolution for systems with multiple electroactive species can be obtained (Osteryoung 1983).

The applied waveform is shown in figure 3.9. A symmetrical square wave is superimposed on a staircase waveform. The period of the square wave is identical to the time step of the staircase. In figure 3.10 the current during one period of the square wave is depicted. Typical values of the pulse width are in the range of a few thousandths of a second to a few tenths of a second. As the pulse duration decreases, more complicated equipment is necessary. The signal current during a SWV experiment equals the difference between the current sampled at the end of the positive pulse (i_2) and the current sampled at the end of the negative pulse (i_1). As the capacitive current decreases much faster than the faradaic current, a mainly faradaic current is measured at the end of the pulse. Also, as the capacitive current is a charging current of the double layer, this current decreases exponentially as a function of time, whereas the faradaic current decreases as a $t^{-1/2}$ function of time, as described by the Cottrell equation (2.31).

A typical SWV experiment is depicted in figure 3.11, showing (*a*) the

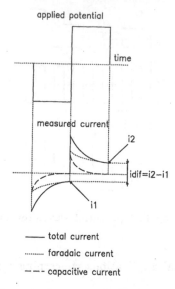

Figure 3.10   The actual current during one period of the square wave.

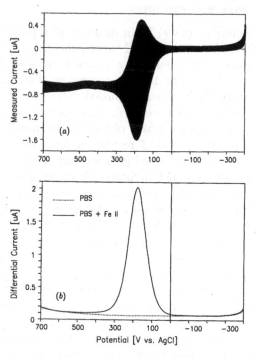

Figure 3.11   Measured current (*a*) and differential current (*b*) of a typical square-wave voltammetry experiment.

combined current i_1 and i_2 and (*b*) the differential current as a function of potential. This latter curve is the square-wave voltammogram. The height of the peak is directly proportional to the concentration of the electroactive species involved. In the experiment of figure 3.11, one electroactive species is present in the solution $[Fe(CN)_6]^{4-}$. An excellent definition of the peak is seen. Therefore, square-wave voltammetry allows the analysis of systems with multiple electroactive species, because the peaks are separated. A SWV experiment is completely characterized by the square-wave frequency (0.1 Hz – 100 Hz), the step height (5 mV – 50 mV), the step increment (1 mV – 5 mV), the initial potential and the final potential. The frequency and the step height have an important influence on peak height and peak separation. By increasing the step height, the sensitivity increases but the peak separation decreases. Hence, a SWV experiment can be optimized for maximum sensitivity or for maximum peak separation.

SWV is mainly used in order to obtain a higher sensitivity. It is also a fast technique. The sensitivity can be further improved by the use of stripping techniques in which the electroactive species involved is concentrated in the working electrode by a suitable pre-polarization of the electrode (Peterson and Wong 1981). With stripping techniques, a detection limit in the 10^{-9} M to 10^{-11} M range can be obtained (Wang 1986). The major disadvantage of SWV is the complexity of the measuring equipment, so SWV techniques are limited to 'in vitro' measurements.

3.2.5   Differential pulse voltammetry

Differential pulse voltammetry (DPV) is also a pulsed voltammetric technique. The applied potential is shown in figure 3.12. Small pulses are superimposed on a linearly increasing potential wave. The pulse duration is typically in the 50 ms region and the pulse height is between 5 and 250 mV. The pulse repetition period can vary from 0.5 s to 5 s. The current is measured twice; once before the pulse and once at the end of the pulse. The signal current is obtained by subtraction of the two measured currents, so the differential pulse voltammogram is, as in SWV, a plot of the differential current versus the applied potential. A faradaic process also results in a peak current in the voltammogram. DPV is generally used in combination with a DME; SWV is also used at solid electrodes. In general SWV techniques are faster than DPV techniques.

3.2.6   Amperometry

Amperometry is a voltammetric technique where the potential is kept constant as a function of time; sometimes this technique is also entitled chronoamperometry. It is the ideal technique for measurements with voltammetric sensors. The measuring equipment is simple and the measured current is a linear function of the concentration of the electroactive

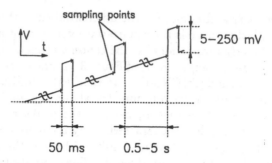

Figure 3.12 The applied potential waveform during differential pulse voltammetry.

species involved. This method is preferred over linear-sweep voltammetry if no potential pre-treatment of the electrode is required.

In section 2.3.1, the Cottrell equation (2.31) was derived for large plane electrodes under semi-infinite linear diffusion conditions. According to this equation the current decreases as $t^{-1/2}$. In practical experiments, however, it is found that, although the Cottrell equation is initially followed, the current reaches a limiting value after approximately 10 to 30 seconds (figure 3.13). This constant current as a function of time is related to two causes. Firstly, the unlimited expansion of the diffusion layer with $t^{1/2}$ is prevented by convection in the solution; after some time the thickness of the diffusion layer becomes constant as a function of time. Secondly, normal electrodes, as used in this work, do not operate under semi-infinite linear diffusion conditions. It can be proved that for spherical or cylindrical electrodes a constant term as a function of time must be added to the Cottrell equation (see equation (3.3) and 3.4). Fortunately, this term is also a linear function of concentration. This implies that, in spite of the deviations of the theoretical model, the limiting current is a linear function of concentration and can be used for analytical applications. Nevertheless, the limiting current at planar microelectrodes can not be predicted with the existing models.

The classical models, as explained in section 2.3.1, are based on the assumption of semi-infinite linear diffusion at large planar electrodes. Because of the small dimensions, these electrodes behave like microelectrodes, especially as the width of the electrode approaches the 10 μm region. Starting from this dimension, these electrodes are named 'ultra-microelectrodes'. It is known from the literature that the characteristics of circular electrodes with a diameter smaller than 10 μm are completely different from those of classical electrodes (Wightman 1981). Since the electrodes are planar and have a finite length and width, edge effects also play an important role (Oldham 1981). The current at microelectrodes is defined by spherical diffusion, not by linear diffusion. This implies that the area of the diffusion

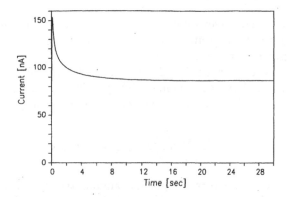

Figure 3.13 Current measured during an amperometric experiment. A constant potential (-700 mV) as a function of time is applied to a planar Au electrode in PBS saturated with air.

layer boundary increases as the thickness of the diffusion layer increases. Therefore, the classical models are not valid any more. In the case of the microelectrodes described in this work (band electrodes), linear as well as spherical diffusion will determine the voltammetric response. Hence, the mathematical solution of this problem is extremely complicated.

However, a simple mathematical solution can be found for classical geometries, such as wire electrodes and spherical electrodes. As already discussed in section 2.3.1, the current during an amperometric experiment under limiting current conditions can be found by solving the boundary value problem.

Under semi-infinite linear diffusion conditions at large electrodes, the current as a function of time is described by the Cottrell equation (Bard and Faulkner 1980):

$$i(t) = \frac{nFAD_O^{1/2}C_O^*}{\pi^{1/2}t^{1/2}}. \tag{3.2}$$

For a wire or cylindrical electrode, the current is defined by cylindrical diffusion and, after solving the boundary value problem, one finds the following equation (Adams 1969):

$$i(t) = nFAD_OC_O^*\frac{1}{r}\left[\frac{1}{(\pi\phi)^{1/2}} + \frac{1}{2} - \frac{1}{4}\left(\frac{\phi}{\pi}\right)^{1/2} + \frac{1}{8}\phi\dots\right] \tag{3.3}$$

where $\phi = D_Ot/r^2$ and r equals the radius of the cylinder. If only the first two terms are taken into account, equation (3.3) is in fact the sum of the Cottrell equation and a constant factor as a function of time. This assumption is valid for small values of ϕ. The constant factor increases as the diameter of the electrode decreases. This factor thus only becomes important for small-diameter electrodes. For long periods of electrolysis, this assumption is no longer valid and the current approaches zero inversely proportionally to the logarithm of time. In practice, convection in the solution will prevent this decay to zero.

A similar solution can be found for spherical electrodes. In this case the current at the electrode is defined by spherical diffusion. Solving this boundary value problem results in the following equation (Adams 1969):

$$i(t) = \frac{nFAD_O^{1/2}C_O^*}{\pi^{1/2}t^{1/2}} + 4\pi r n F D_O C_O^*. \tag{3.4}$$

In this case too the Cottrell equation augmented by a constant factor proportional to the concentration is found. For a smaller electrode radius r, the constant factor increases.

The problem has also been studied recently for microband electrodes by Aoki *et al* (1987a). He found a very complex expression for the current at this type of electrode. Fortunately, he also reported an approximate solution that corresponds within 1% to his theoretical solution (Aoki *et al* 1987b). This solution is a function of the dimensionless parameter θ; the width of the electrode equals w, the length equals b:

$$q = D_O t / w^2. \tag{3.5}$$

The current at microband electrodes can then be expressed by the following approximate equation:

$$i(t) = nFD_O bC_O^*((\pi\theta)^{-1/2} + 0.97 - 1.10 \exp[-9.90/|\ln(12.37\theta)|]). \tag{3.6}$$

In figure 3.14, these different models are compared with each other and with an amperometric measurement at a 'VOLTA II' Au electrode (i.e. an electrode configuration developed by the authors, see further). The width of this electrode is 10 μm. From figure 3.14, it is clear that the normal Cottrell equation is not followed at all. It appears that the response of the Au electrode equals the Cottrell equation with a constant term added. The same response is seen for the spherical and cylindrical model. For the spherical model a sphere with the same area as the band electrode is assumed. For the cylindrical model, half of the Cottrell current and half of the constant current are taken from a wire electrode with the same diameter as the width of the band electrode, since diffusion takes place at only one side of the band electrode.

The shape of the current response of the spherical and the cylindrical models corresponds well with the shape of the measured response. The deviation in absolute value is not relevant as the diffusion coefficient of oxygen in PBS and the exact oxygen concentration cannot be known precisely. A variation of about 20% can therefore be expected. The shape

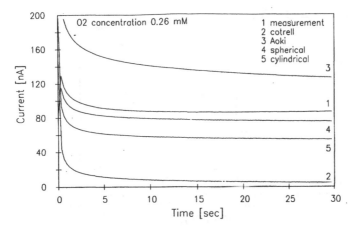

Figure 3.14 Chronoamperometric response of a 10 μm wide Au electrode, compared with the different models for microelectrodes.

of the response according to the model proposed by Aoki does not match satisfactorily the measured response. This is in contradiction to the measurements described in Aoki *et al* (1987b); this deviation can be related to a larger range of the θ parameter in our experiments in comparison with the values reported in Aoki *et al* (1987b). Based on this brief comparison of the different available models with a typical measurement on a band microelectrode, it can be concluded that the response of such an electrode approximates the response of a wire electrode or a spherical electrode, and that a Cottrell relationship is found combined with a constant term.

Therefore, new techniques and models are necessary for the prediction of the response of planar microelectrodes. As a novel technique for the simulation of planar electrodes, the finite-element method is proposed (Lambrechts 1989a). Introductory results indicate that the current at an electrode of arbitrary shape can be calculated with finite-element analysis. However, further investigation is still necessary.

An important advantage of amperometry is that it is a fast technique. In figure 3.15, ten points on a calibration curve of a Pt H_2O_2 detector are measured in 10 minutes by the injection of a concentrated solution every minute. If the same measurement has to be carried out with linear-sweep voltammetry, at least a few hours of experiment time are necessary to obtain equivalent information. The disadvantage of amperometry is that only one point of the complete current–potential curve is followed as a function of time. If the electrode becomes poisoned or the package is damaged, no distinction can be made between these malfunctions or a change in concentration.

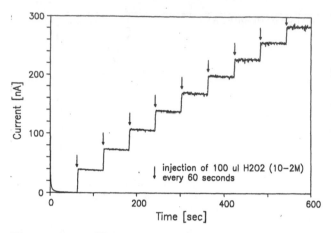

Figure 3.15   The current as a function of time at a planar Pt electrode for different stirred PBS–H_2O_2 concentrations during an amperometric experiment.

3.2.7   AC impedance techniques

AC impedance measurements are not really voltammetric experiments, although they are also based on the application of a potential and the measurement of the corresponding current. This technique can, however, provide a wealth of kinetic and electrode information. This method is of special
interest for the study of membrane, electrode and electrolyte characteristics. AC impedance measurements are not used for analytical applications.
Conductometry (i.e. the measurement of the resistivity), on the other hand,
is used frequently for analytical applications.

The final goal of AC impedance measurements is to find an equivalent
electronic model and to correlate that model with electrochemical phenomena. An equivalent electronic circuit for a simple electrochemical electrode
is shown in figure 3.16. It consists of R_Ω (the uncompensated solution resistance), C_d (the double-layer capacitance) and R_{ct} (the charge-transfer
resistance). If a faradaic process occurs, limited by diffusion, a Warburg
impedance Z_W can be placed in series with the charge-transfer resistance
(Bard and Faulkner 1980). All these components can then be related to
electrochemical processes taking place in the double layer (figure 3.17).

The Warburg impedance, Z_W, represents a kind of resistance to mass
transfer; it is not an ideal component, as it changes with frequency. For a
diffusion-controlled faradaic reaction, the phase of the current is shifted 45°
with respect to the applied potential in the Warburg impedance. The Warburg impedance can be represented as a frequency-dependent resistance
$R_W = \sigma/\omega^{1/2}$ in series with a pseudo-capacitance $C_W = 1/(\sigma\omega^{1/2})$. The
total Warburg impedance, $|Z_W|$, varies with frequency as $\omega^{-1/2}$. This also

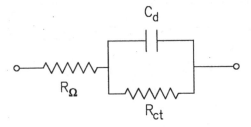

Figure 3.16 Equivalent electronic circuit for a simple electrochemical process.

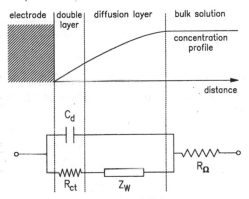

Figure 3.17 Correlation of the different equivalent components of a faradaic process with electrochemical phenomena.

implies that the real and imaginary components of the Warburg impedance are equal, independent of the frequency. The Warburg impedance becomes predominant at low frequencies.

For AC impedance measurements, very small excitation amplitudes (5 to 10 mV peak-to-peak) are used. This sinusoidal potential is applied to the electrochemical cell and the resulting AC current is measured. As one wants to measure the impedance of only one electrode, a two-electrode system with a large reference electrode or a three-electrode system has to be used. An AC impedance measurement now consists of the measurement of the impedance for different frequencies. The frequency range of interest for electrochemical experiments begins at 1 mHz and stops at 20 kHz.

The impedance information as a function of frequency can be represented in different ways. The best known representation for electronic engineers is the Bode plot. The absolute impedance of the electrode, $|Z|$, and the phase-angle, θ, are plotted as a function of frequency. In figure 3.18, a simulated Bode plot is shown for the simple electrochemical cell depicted in figure 3.16. All the components can be derived graphically from this representation.

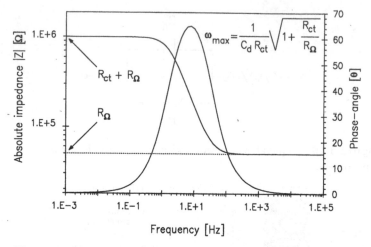

Figure 3.18   Simulated AC impedance measurement Bode plot of the simple circuit shown in figure 3.10 ($R_\Omega = 50$ kΩ, $R_{ct} = 950$ kΩ, $C_d = 100$ nF).

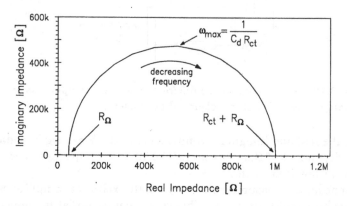

Figure 3.19   Simulated AC impedance measurement Nyquist plot of the simple circuit shown in figure 3.10 ($R_\Omega = 50$ kΩ, $R_{ct} = 950$ kΩ, $C_d = 100$ nF).

Electrochemists prefer the Nyquist representation or polar diagram of the impedance, in which the negative imaginary impedance is plotted. This format is also known as a Cole–Cole plot or a complex impedance plane diagram. The imaginary component of impedance is plotted versus the real component of impedance. In figure 3.19 a simulated Nyquist plot is shown for the simple electrochemical cell depicted in figure 3.16. In this representation, all the components can also be derived graphically. For this simple circuit the Nyquist diagram results in a half-circle. The Nyquist plot is of special benefit for more complicated systems, such as, for example, a

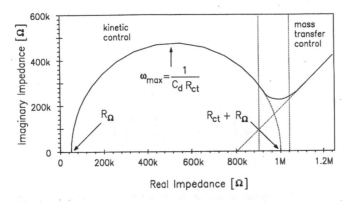

Figure 3.20 The effect of a Warburg impedance on an AC impedance measurement Nyquist plot.

Warburg impedance or multiple-membrane systems. In figure 3.20 the effect of a Warburg impedance on a Nyquist diagram is shown. Under kinetic control (high frequencies) a half-circle is seen; under mass transfer control (low frequencies) the Nyquist diagram is reduced to a 45° line.

In this work, AC impedance measurements are not used. Nevertheless, it is a very powerful tool for the evaluation of electrodes and membrane systems.

3.3   CONDUCTOMETRIC TECHNIQUES

Conductometric techniques are in fact a special case of AC impedance techniques. Instead of the real and imaginary component of the electrode impedance at different frequencies, only the resistive component, related to the solution resistance, is of interest. The measurement is carried out at one fixed frequency. As discussed in section 2.4, the point of interest of conductometry is situated in the bulk of the solution; interference of the electrode impedance has to be avoided.

Therefore, a four-electrode configuration is preferred for more precise measurements. As depicted in figure 3.21, a four-electrode conductometric cell consists of two pairs of electrodes. At the outer pair, an AC current is injected into the solution; the potential difference is measured at the inner electrodes. In this way, interference of the electrode impedance is avoided and more precise measurements can be carried out. An increase in sensor complexity, however, limits the use of four-electrode configurations if the available space is a limitation.

The equipment for conductometric measurements can be kept very simple. Before the existence of modern electronic equipment, the conductance was measured with a Wheatstone bridge. A better solution is shown in

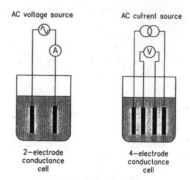

Figure 3.21   Measurement set-up for two-electrode and four-electrode conductometric cells.

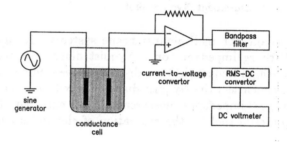

Figure 3.22   Conductometric measurement set-up based on a RMS-DC convertor.

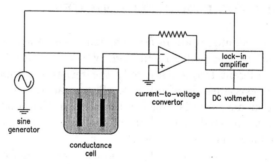

Figure 3.23   Conductometric measurement set-up based on a lock-in amplifier.

figure 3.22. An AC voltage with a small amplitude is applied to the conductance cell, resulting in an AC current. This is converted to a voltage in a current-to-voltage convertor. The resulting AC voltage is then filtered in a bandpass filter and transformed to a DC voltage in a RMS-DC convertor, which can now be measured with a simple DC voltmeter and is directly proportional to the conductance of the cell.

More accurate measurements can be performed with a lock-in amplifier. The basic measurement set-up is shown in figure 3.23. The lock-in amplifier measures the amplitude of the AC signal, for a particular phase characteristic and frequency of the applied signal.

3.4 CONCLUSION

Evaluation of microelectrochemical sensors is of prime importance in the development phase of these sensors. As chemical sensor measurements are time consuming, it is advisable to have a computer-controlled measurement set-up for the evaluation.

In this chapter, electrochemical techniques and the corresponding equipment are reviewed for measurement, and the evaluation of microelectrochemical sensors. Special attention is paid to potentiometry, voltammetry and conductometry. It is worth noticing that modern electronic circuits become more and more important for these applications. Cheap pH meters, as well as expensive computer-controlled potentiostats, make extensive use of electronic circuits. For microelectrochemical devices, high-precision equipment is necessary, since the current decreases with the area of the electrodes. Hence, the next step is the integration of the interface electronics together with the sensor on one substrate, resulting in a smart sensor. Examples of smart sensors are given in section 5.3.1 and 5.3.4.

CHAPTER 4

PLANAR TECHNOLOGIES FOR MICROELECTROCHEMICAL SENSORS

Different technological aspects of microelectrochemical sensors are discussed in this chapter. Two basic microelectronic fabrication technologies can be used for the realization of interface circuits of planar sensors, namely CMOS and bipolar technology. Integration of the sensor on the interface circuit chip is the ultimate goal of planar smart sensors. Therefore, compatibility of the sensor process with the interface circuit technology is of crucial importance in the development of the sensor process.

A bipolar or a CMOS (complementary metal oxide semiconductor) process is the ideal technological basis for microelectrochemical sensors. This technological basis, combined with different special sensor techniques, such as lift-off, micromachining in silicon and membrane formation technology, results in a generic CMOS-compatible sensor process. The different techniques are detailed in this chapter. Of course, special attention has been focused on sensor applications of the different processing steps. The realization of the sensor itself with the techniques described in this chapter is detailed in chapter 5.

Thick-film technology is also proposed as an alternative to microelectronic fabrication techniques. Thick-film technology is based on screen printing and firing of conductive, resistive and insulating pastes. In comparison with a thin-film based process, screen-printing techniques have the advantage that all pastes can be printed and fired with the same equipment. This results in a low investment and production cost per sensor. However, because of the larger dimensions, thick-film sensors are more suited for 'in vitro' measurements. Thick-film technology is explained in section 4.5. The realization of thick-film sensors will be discussed in chapter 6.

When CMOS and thick-film technology are explored, a comparison between both techniques reveals that CMOS-compatible sensors are ideal for 'in vivo' applications, whereas thick-film sensors are better adapted to 'in vitro' applications. Therefore, a complete range of microelectrochemical sensors can be made with these two sensor technologies.

4.1 STANDARD MICROELECTRONIC FABRICATION TECHNOLOGY

Essential technological components for planar electrochemical sensors with on-chip interface electronics are:

- planar, noble-metal electrodes
- high-quality insulation layers between the electrodes
- a corrosion-free, well insulated interconnection layer
- a fluid-resistant passivation layer
- operational amplifiers with high input impedance.

For biomedical applications, such as an implantable glucose or oxygen sensor, special requirements are:

- small size
- low power consumption
- HF-surgery and defibrillation resistant electronics.

Practically all these requirements can be implemented in standard CMOS or bipolar technology. In comparison with bipolar technology, CMOS technology has the advantage that high-impedance operational amplifiers are directly available. In a bipolar process, addition of JFETs provides the same advantage. Also, with switched-capacitor circuits, reliable and accurate filters and current-to-voltage converters can be developed. Switched-capacitor techniques are not easy in a bipolar process. On the other hand, it is generally assumed that bipolar technology is less sensitive to static discharge and defibrillation pulses.

So, both technologies have distinctive advantages as a technology base for smart chemical sensors. There are also economic factors, such as availability, standardization and price, that influence the choice between a bipolar and a CMOS process. An interesting new development in the field of circuit technology is the increasing availability of BICMOS (BIPOLAR-CMOS) processes. In a BICMOS process CMOS transistors, as well as bipolar transistors, are available. In this way the advantages of both technologies can be exploited. No realizations of a chemical sensor in combination with a BICMOS circuit are known to the authors. The specific advantages of BICMOS for analogue as well as digital circuits (Pfleiderer *et al* 1989) however, allow us to presume that interesting combinations can be found for microelectrochemical sensors.

4.1.1   A standard CMOS process

CMOS (complementary metal oxide semiconductor) technology is a wide-spread production technique for high-performance, digital and analogue integrated circuits. In contrast with NMOS (n-channel) technology and PMOS (p-channel) technology, CMOS is based on both p-channel and n-channel MOS transistors. It offers interesting properties such as high packing density, high input impedance, low-voltage power supply and low power consumption. These properties ensure that CMOS technology combines very well with planar chemical sensors.

A standard CMOS process can be divided into the following basic steps:

1. cleaning of the wafers (p-type or n-type)
2. delineation of active areas with CVD silicon nitride
3. n-well or p-well implantation and drive-in
4. field oxidation
5. gate oxidation
6. polysilicon deposition, doping and etching
7. p^+ source and drain implantation
8. n^+ source and drain implantation
9. deposition of CVD insulation oxide
10. etching of contact holes
11. aluminium deposition, etching and sintering
12. passivation
13. opening of bonding pads.

The complete cross section of the final structure is shown in figure 4.1. This process sequence is almost identical for all CMOS silicon foundries, for n-well as well as for p-well technology. So, a sensor process compatible with this sequence can be applied to processed CMOS wafers of different suppliers.

4.1.2   A standard bipolar process

Bipolar technology is the oldest production technique for integrated circuits. In figure 4.2 the cross section of a typical bipolar process, known as the standard buried-collector process, is shown. The buried layer, or subcollector, is introduced in order to reduce the collector resistance. The different adjoining devices are separated by back-to-back pn diodes realized by the isolation diffusion. In more sophisticated bipolar processes, oxide or trench isolation is used to increase the performance of the devices.

The standard-buried collector process can be divided into the following basic steps:

1. cleaning of the wafers (p-type)
2. diffusion of the buried layer (n^+)

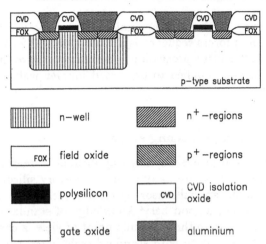

Figure 4.1 The cross section of a CMOS structure.

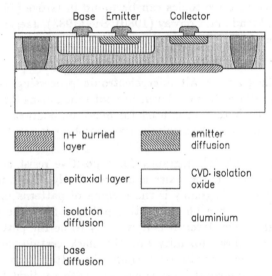

Figure 4.2 The cross section of the standard buried collector bipolar process

3. epitaxial layer growth (n-type)
4. isolation diffusion (p)
5. base diffusion (p)
6. emitter diffusion (n)
7. etching of contact holes
8. aluminium deposition, etching and sintering
9. CVD oxide passivation
10. opening of bonding pads.

The complete cross section of the final bipolar structure is shown in figure 4.2. This basic process sequence is almost identical for most bipolar silicon foundries. As for CMOS processing, a sensor process compatible with this sequence can also be applied to processed bipolar wafers of different suppliers.

4.1.3   Sensor-related processing steps

In this section, standard processing steps applicable to sensor processing are detailed. All these steps are daily practice in every silicon processing area. To understand the special sensor processing techniques discussed in sections 4.2, 4.3 and 4.4, a good basic knowledge of standard processing steps is necessary. Also, from the economic point of view, a cheap, reliable sensor process must be built up with standard techniques, where possible. Non-standard processing is expensive and slow.

A good coverage of these topics can be found in Grove (1967), Vossen and Kern (1978), Mead and Conway (1980), Sze (1983), Jaeger (1988) and other standard text books on microelectronics.

4.1.3.1 Photolithography. All microelectronic processing is based on photolithography; all the different layers are patterned using this technique. Nowadays, positive photoresist is used for most applications. As will be discussed in section 4.4.3, chemically active membranes can be converted to negative resist.

The basic steps of photolithography for a positive resist are shown in figure 4.3. In this example the wafer is covered firstly with a metal layer. The purpose of photolithography is the etching of patterns in this layer. The wafer is coated with a photosensitive polymer film (the photoresist layer) on top of the metal layer; it is now ready to be the patterned. The wafer is brought into close proximity with the mask, certain parts of which are covered with an opaque material, forming the pattern to be transferred to the metal layer. The resist layer is exposed with UV light through the openings on the mask; at the opaque regions, this UV light is stopped. In the exposed regions, the molecular structure of the organic resist material is altered; the solubility in a suitable solvent is higher, so that the resist in the exposed regions is removed during development in the solvent.

Thus far, the pattern on the mask has been transferred onto the resist layer. Etching of the metal layer is the next step. Etching is carried out by exposing the wafer to a liquid that will etch the metal layer but leave the resist layer and the substrate intact. The final step in photolithography is the removal of the remaining resist material. In this way, patterns on a mask can be transferred to a great variety of materials: Si, SiO_2, Si_3N_4, aluminium and other metals, ...

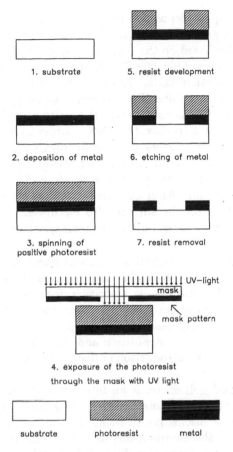

1. substrate

5. resist development

2. deposition of metal

6. etching of metal

3. spinning of positive photoresist

7. resist removal

4. exposure of the photoresist through the mask with UV light

UV–light

mask

mask pattern

substrate   photoresist   metal

Figure 4.3   The basic steps of photolithography.

It is important to notice that a resolution of 5 μm is adequate for sensor processing. A simple contact aligner with optical masks will do the job. There is no need for high-priced wafer steppers or expensive electron-beam masks.

4.1.3.2 Thermal and CVD oxides.   Silicon dioxide is used as an insulating layer. Different quality is obtained depending on the processing method. Dry oxidation of the silicon wafer in a pure oxygen atmosphere produces the best quality oxide. This thermal oxide is stoichiometric, has a high density and is basically pinhole free. In CMOS processing, it is used for the gate oxide.

For thick oxides (e.g. a field oxide of 1.2 μm) wet oxidation in steam is used, but the quality of wet oxides is somewhat less than for dry oxides. Both types of thermal oxides are grown in a quartz tube at temperatures ranging from 900 °C to 1150 °C.

Chemical vapour deposition (CVD) is an alternative method. CVD silicon dioxide is not grown but deposited on the wafer out of the vapour phase. This method is based on the reaction of silane (SiH_4) with oxygen at moderate temperatures (300–500 °C). In a CMOS process, this type of oxide is used as the isolation oxide. To obtain a low melting point of the glass a few per cents of phosphine (PH_3) is added to the gas stream during deposition. In this case a phosphosilicate glass (PSG) is realized. The low melting point is necessary to re-flow the glass in order to increase the density. In a bipolar process, CVD oxides are used for passivation.

The CVD method results in a rather low-quality, porous oxide layer. It is a poor insulator for chemical sensors, since the oxide reacts as a sponge in aqueous solutions. After extended exposure to water, a conducting hydrogel is formed on the surface. To circumvent this problem, an extra passivation layer of silicon nitride is necessary (see section 4.1.3).

4.1.3.3 Polysilicon.

Polysilicon, a short name for polycrystalline silicon, is the gate material in a modern MOS process. It is deposited in a LPCVD (low pressure chemical vapour deposition) furnace at temperatures of 600°C to 750°C. The typical gas pressure is 25 to 150 Pa. Under these conditions, silane gas is thermally decomposed into polycrystalline silicon. These films are made conductive by deposition and drive-in of phosphorus, by ion implantation or by the introduction of dopants in the gas stream of the LPCVD furnace.

Highly doped polysilicon is an interesting interconnection material for sensor applications. In comparison with diffused layers, a polysilicon interconnection is insensitive to light. It is corrosion resistant and is well protected from the environment, as it is encapsulated within two silicon dioxide layers (see figure 4.1). Due to the rough surface, adhesion of metal layers is better; ohmic contacts are easily made to this highly doped material.

Recent interest in polysilicon as sensor material is based on the exciting piezoresistive and temperature characteristics of this material (Obermeier 1986, Guckel *et al* 1987). It is also an important material in the so-called surface-machining or sacrificial-layer technique (Muller 1989). Although these properties are not directly applicable to chemical sensors, this research proves that a great variety of all types of integrated sensors based on polysilicon will soon emerge.

4.1.3.4 Evaporation and sputtering.

To apply thin metal films (e.g. aluminium) on a substrate, evaporation and sputtering techniques are widely used in semiconductor processing facilities. Evaporation techniques are based on the evaporation of the heated metal in high-vacuum conditions. Different heating methods are known. The metal can be evaporated by passing a high current through a filament or a boat; this method is known

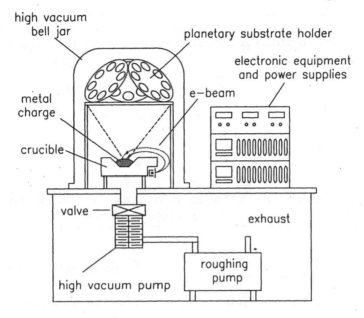

Figure 4.4   Schematic representation of an electron-gun evaporator

as 'resistive heating'. Resistive heating has been replaced by electron-beam heating for most applications (laboratory experiments excepted), so in modern evaporation equipment an electron-beam gun is installed to heat up the metal. A high-intensity electron beam with an energy up to 15 keV is focused on the target material, which, due to the high energy of the electron beam, melts and starts to evaporate. The target material is placed in a multi-crucible with several holes so that thin-film multilayers of different metals (e.g. Ti/Pd/Ag) are easily produced. A schematic diagram of an electron-gun evaporator is shown in figure 4.4. To increase the uniformity of the deposited layer, the wafers are mounted on a planetary substrate holder; this planetary substrate holder rotates simultaneously around two axes.

Sputtering is preferred over evaporation in most applications due to the difficulties in evaporating alloys (e.g. Al with 1% Si), a higher deposition rate and better step coverage, adhesion and purity of sputtered thin films.

In a sputter process, the target (a plate of the material to be deposited) is bombarded with positive argon ions in an argon plasma. The material is sputtered away from the target by momentum transfer and deposited on the substrate (figure 4.5). The target is connected to a negative RF or DC voltage supply (a DC power source is used for electrically conductive materials). For the sputtering of dielectric material, an RF power source is required. The argon is introduced into the vacuum chamber at a certain flow rate so that a well defined sputter process is maintained. The typical

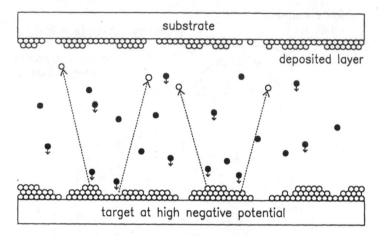

positive ions accelerated from   ⚲ ejected surface atom
plasma ion cloud to target

Figure 4.5   The sputtering mechanism based on momentum transfer.

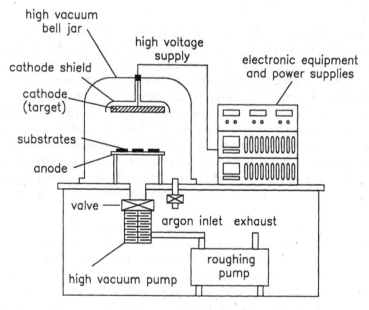

Figure 4.6   Schematic representation of a sputtering machine.

gas pressure ranges from 10^{-3} mbar to about 10^{-1} mbar. A schematic of a sputtering machine is shown in figure 4.6.

As there is a large set of process parameters (sputtering pressure, base pressure, substrate temperature, sputter power, bias voltage, electrode distance, gas flow and sputtering mode), a great variety of specific sputtering processes with well defined characteristics can be produced in the same sputtering machine. This variety also implies that a sputter process is more complex than an evaporation process.

For sensor applications, sputtering is preferred over evaporation. The material is deposited with high mechanical energy during sputtering, so adhesion of the sputtered layer to the substrate is better. To increase adhesion, the substrate can be sputter-cleaned or sputter-etched before deposition. For this purpose, the substrate is connected to the negative power supply instead of the target, so that the substrate is cleaned with bombarding argon ions. With this method, the native oxide on silicon can also be removed in situ before sputtering.

The purity of a sputtered layer is also higher than for an evaporated layer. The grain structure, the surface quality and the density of the sputtered thin films can be influenced by the different parameters (Thornton 1975 and Westwood 1974). To develop an ideal sputter process for thin-film Pt electrodes, a good adhesion, the same density as bulk Pt, high purity, high optical reflectance and a smooth surface have to be obtained. This can be achieved with a low sputter pressure, a high sputter power and a high substrate temperature. However, a high substrate temperature is in conflict with lift-off patterning techniques, as will be discussed in section 4.2. A good compromise has to be chosen.

4.1.3.5 Passivation. Since microelectrochemical sensors are exposed to aggressive and hostile environments, an excellent passivation of the electronic circuits and the interconnections is essential for the long-term activity of the sensor. Because protection from the aggressive environment is also a packaging problem, passivation and packaging of the sensor have to be harmonized. Passivation is typically done at wafer level, whereas packaging is a single die process for an individual sensor. In this section, passivation is discussed. The packaging of chemical sensors and associated problems are described in section 5.1.3.

In semiconductor processing, three different types of passivation layers are in use: silicon dioxide (SiO_2), silicon nitride (Si_3N_4) and organic or polymer films (Schnable *et al* 1975). For experimental devices aluminium oxide (Al_2O_3) films are sometimes used.

SiO_2 films do not provide effective passivation for an aqueous or humid environment, since the silicon dioxide hydrates, forming a conductive hydrogel. Low-temperature CVD techniques (LTCVD), used for passivation purposes, also result in a porous film with pinholes. Modern plasma-enhanced CVD techniques (PECVD) solve these problems so that dense, homogeneous films can be produced. However, this kind of passivation layer only gives scratch protection for chemical sensors because of the hydratation problem.

Si$_3$N$_4$ films provide excellent passivation, especially high-temperature, low-pressure CVD silicon nitride films (LPCVD), which are known to be dense and stoichiometric with zero water permeability (Kern and Rosier 1977). These films, however, are not suited to the passivation of CMOS devices because of the high process temperature (900°C). Therefore, PECVD nitride films are used for passivation layers in CMOS processing. If deposited under optimum conditions, PECVD nitride films are practically pinhole free, scratch resistant, have a good conformal step coverage and are an effective barrier to alkali ions. The chemical resistance of this material is also high. The water permeability is not zero but is still very low. A long-term immersion test, in physiological solution, of plasma silicon nitride on silicon revealed an etch rate of 0.1 nm per day. The result of this experiment indicates that for very long-term implantation (> 2 years) even plasma Si$_3$N$_4$ passivation is not sufficient for protection against a saline environment.

Polymer or organic films are relatively interesting passivation materials, as they are easily processed with low-cost equipment. Polyimide films have the best reputation for passivation. Three patterning techniques are in use. The first method, based on the wet etching of the half-cured polyimide during development of the top photoresist layer, gives a poor resolution; the reproducibility of the process is also bad. Dry etching of polyimide in an oxygen plasma gives very good results. This method, however, requires a metal mask in most cases, as the etch rate of photoresist is higher than the etch rate of polyimide. After etching the polyimide in an oxygen plasma, a carbon residue is found on the surface, which can influence the electrochemical characteristics of the sensor. For sensor applications, a photosensitive polyimide is preferred, as this technique does not require a costly plasma etcher. Photo-sensitive polyimide is now available from different sources; e.g. Toray, Ciba Geigy and Dupont.

Another polymer, 'Parylene' from Union Carbide, is sometimes used as a passivation layer. This polymer is prepared with a special vapour-hase deposition method and results in a thin, pinhole-free conformal coating. Since this method is mostly applied to individual sensors or circuits and not on the complete wafer, this method will be discussed in the section on packaging (see section 5.1.3).

4.2   LIFT-OFF TECHNIQUES

In this and following sections, attention is paid to special sensor technology. These sensor techniques have been developed by the authors and are derived from standard processing techniques or from experimental research. They are necessary to apply the non-standard materials, such as noble-metal layers, chemically sensitive polymers and biological material, on the silicon structure.

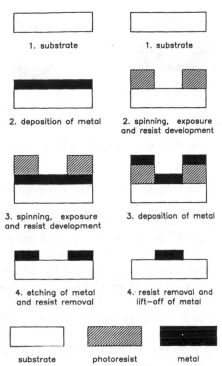

Figure 4.7   The basic principle of lift-off compared with etching.

Lift-off is an alternative to conventional etching techniques for the patterning of metal layers. The lift-off method involves the formation of a patterned photoresist layer prior to deposition of the layer for which patterning is desired. By subsequent dissolution of the photoresist, and thereby removal of the metal layer on top of this photoresist, the metal pattern is formed. As depicted in figure 4.7, the lift-off method is based on the swapping of the photolithographic step with the metal deposition step. It is also important to notice that with lift-off the reverse structure is obtained compared to etching. This implies that, in order to obtain the same metal pattern, the photomask for lift-off must have the inverse field in comparison with the photomask for etching.

Lift-off techniques offer better line-width control in the micron and submicron region due to the fact that the dimensions of the metal pattern are not defined by the resolution of the etching process, but are determined by the resolution of the positive photoresist layer. Lift-off also allows the patterning of metallizations which are difficult or impossible to etch. Good examples of materials which are difficult are gold, platinum and especially multi-layers such as Ti/Pd/Ag, Cr/Au or Ti/Pd/Au. Since these noble-metal layers are intrinsic components of electrochemical sensors, lift-off

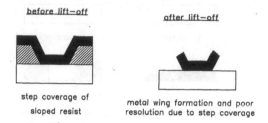

Figure 4.8   Step coverage of sloped resist and the effect on lift-off behaviour.

techniques are of critical importance in the development of planar electrochemical sensors.

The use of lift-off methods also gives the advantage that the patterning process is identical for all metal layers, so that different electrode materials can be evaluated with the same technique. For example, the following electrode materials have been patterned by the authors by the lift-off method described below: Ag, Ti/Ag, Cr/Ag, Ti/Pd/Ag, Ti/Pd/Ag/AgCl, Au, TiW/Au, Ti/Pd/Au, Pt, Ti/Pd/Pt, C, Al and Ti/Al. The optimization of etch techniques for all these metallizations would have taken several months.

In this book two different lift-off methods developed by the authors are described: a tri-level lift-off method and a CVD oxide assisted lift-off method.

4.2.1   Lift-off methods

A simple lift-off method, as depicted in figure 4.7, includes the following steps:

1. cleaning of the substrate
2. spinning and pre-bake of the photoresist
3. development of the photoresist
4. sputtering or evaporation of the metal layer
5. dissolution of the photoresist in acetone and lift-off of the metal layer with ultrasonic agitation.

This elementary lift-off method assumes that the resist profile is ideal with a 90° slope and that the deposition method has a zero-step coverage. Real life, however, is not that simple. This method only works with very thin, highly stressed, brittle layers, as the resist slopes are covered with the metal film. Indeed, the cross-sectional profile of a resist opening in a standard process is larger at the top than at the bottom. In this case, the lift-off of the metal layer becomes difficult because of the inhomogeneous step coverage. This results in poor edge definition and the formation of metal wings at the edge of the metal lines (figure 4.8).

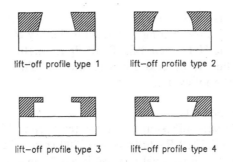

lift—off profile type 1   lift—off profile type 2

lift—off profile type 3   lift—off profile type 4

Figure 4.9   Different lift-off resist profiles with reverse taper.

To obtain good line-width control with thick metal layers, a reverse-taper profile (with a smaller opening at the top than at the bottom) is necessary to prevent step coverage. A great variety of reverse-taper resist profiles can be obtained with different methods; all the structures in figure 4.9 prevent step coverage, so that lift-off of the subsequent metal layer can be performed.

Different lift-off techniques are described extensively in (Frary and Seese 1981). Three main classes of lift-off methods can be distinguished according to the total number of layers necessary to obtain the desired resist profile: single layer, bi-layer and tri-layer lift-off methods. The best known lift-off technique is the 'chlorobenzene lip process' described in (Hatzakis *et al* 1980). This single layer lift-off method for Shipley's AZ1300, AZ1400 and AZ2400 series is based on the hardening of the top of the resist layer by a chlorobenzene soak after the pre-bake step. The chlorobenzene removes some of the low molecular-weight components of the resist so that the top layer becomes less soluble during development and that an overhanging lip is formed. This methods typically results in a type 3 resist profile (figure 4.9). Important drawbacks of the chlorobenzene technique, however, are the toxic and carcinogenic properties of chlorobenzene, so this method should be avoided whenever possible. The reproducibility of the process is also questionable.

Bi-layer lift-off methods are based on the combination of two resist layers with different solubilities. To form a structure with overhang, a less soluble resist layer is spun on top of a more soluble resist. Without special treatment of the first resist layer, a mixing of the layers occurs, resulting in a type 2 profile (figure 4.9). If mixing of the two layers is prevented by, for example, a CF_4/O_2 plasma treatment of the first layer or by correct choice of the resist types, type 3 or type 4 profiles can be obtained.

4.2.2 A tri-level lift-off method

A very simple tri-level lift-off method is described in (Grebe *et al* 1974). A tri-level lift-off process has been developed by the authors in cooperation with L Stevens (K U Leuven). This process has been derived from a high-resolution resist-patterning method for the dry etching of polysilicon, developed at the K U Leuven–ESAT. In figure 4.10, the schematic cross section of the different process steps is shown. The rather complicated process involves the following:

1. Cleaning of the wafer.
2. Spinning of HMDS adhesion promoter (5000 rpm, 30 s).
3. Spinning of AZ 1350 H photoresist (5000 rpm, 30 s).
The first layer of this tri-level process is a 1.3 μm thick resist layer.
4. Baking of the resist at 120°C in ambient N_2 (30 min).
The bake temperature is a critical parameter in this process as this temperature will define the temperature stability of the lift-off profile during further processing and sputtering. A higher bake temperature would increase the stability of the profile, but would also prevent the removal of the resist in acetone during the lift-off step (14).
5. Evaporation of 100 nm aluminium.
The second layer of this tri-level process is a 100 nm thick aluminium layer. This layer will form the overhang of the reverse-taper lift-off profile.
6. Spinning of HMDS adhesion promoter (5000 rpm, 30 s).
7. Spinning of AZ 1350 photoresist (2000 rpm, 30 s).
The third layer of this tri-level process is a 0.6 μm thick photoresist layer. The AZ 1350 resist has a lower viscosity and a higher sensitivity than the AZ 1350 H resist. Since the lower viscosity results in a thin resist layer, a very high resolution is obtained (1 μm). The standard AZ 1350 H resist can be used if this high resolution is not required.
8. Pre-bake of the resist at 94 °C in ambient N_2 (30 min).
9. Exposure in hard contact (9 s at 5 mW cm^2).
10. Resist development (Microposit 351, 60 s).
11. Post-bake of the resist at 94°C in ambient N_2 (30 min).
Steps 7 to 10 are standard for a photoresist process.
12. Wet etching of the Al layer.
13. Planar etch in oxygen plasma (30 mTorr, 250 W, 20 min).
The oxygen plasma removes the AZ 1350 resist layer on top of the aluminium and etches the AZ 1350 H resist layer. Due to the isotropic characteristics of the planar etch method and due to a well timed over-etch of the bottom resist layer, an ideal type 3 lift-off profile is obtained. The RF power supplied to the plasma is a critical parameter here, as the wafer may not heat up above 100°C. At high temperatures, the bottom resist layer will deform and small amounts of solvent in this resist layer will be released. This results in the formation of blisters in the aluminium layer and the destruction of the lift-off profile.
14. Evaporation or sputtering of the metal.
Wafer temperature and the sputter power are also of critical importance

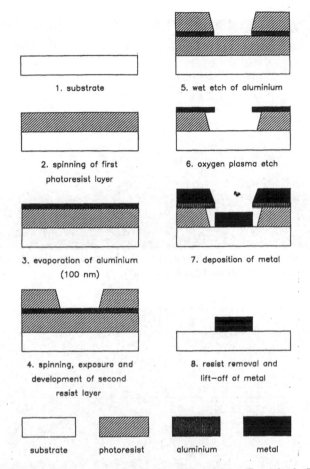

Figure 4.10   The process flow of the tri-level lift-off method. The cross section of the profile is shown

here. A wafer temperature above 100°C will destroy the lift-off profile.

15. Resist removal and lift-off of the metal in acetone at room temperature with ultrasonic agitation.

The time required to lift the metal over the complete wafer depends on the pattern on the mask. For a light-field mask, lift-off is accomplished in less than 60 seconds. For a dark-field mask, or for masks where large areas of metal have to be lifted, the lift-off time may extend up to 15 minutes.

This method gives very good results. In figure 4.11 an SEM photograph of the lift-off profile before metal deposition is shown. The aluminium overhang that provides the reverse taper or 'mushroom' profile can clearly be distinguished. The overhang is approximately 0.6 μm; this overhang, in combination with a thickness of 1.3 μm of the first resist layer, is sufficient

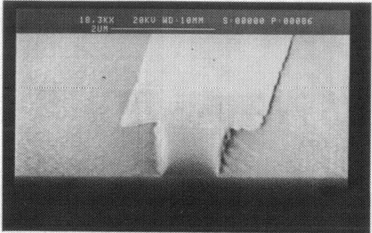

Figure 4.11 SEM photograph of the cross section before metal evaporation (tri-level lift-off method).

Figure 4.12 Microphotograph of 0.5 μm thick aluminium pattern with 2 μm resolution.

to lift metal layers as thick as 1 μm. The following metal layers have been patterned with this technique: evaporated Al and sputtered Ag, Au, Pt and Cu. No problems were encountered with the evaporated samples, although the evaporation was not perpendicular to the substrate. In figure 4.12, a patterned 500 nm thick aluminium film is shown. It can be concluded from the test structures that a resolution better than 2 μm has been achieved. This resolution is more than sufficient for sensor applications.

With the sputtered samples, a low sputter power has been used to avoid high wafer temperatures. At temperatures above 100°C, the bottom resist

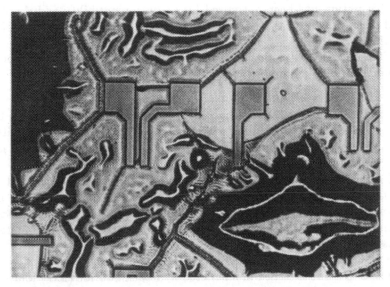

Figure 4.13   A destroyed lift-off resist pattern after sputtering at high power.

layer starts to flow. This results in the destruction of the lift-off profile, so that lifting of the metal becomes difficult. At higher temperatures the solvents in the resist start to evaporate so that blisters in the metal layer are formed. This results in a complete devastation of the resist pattern. In figure 4.13 the microphotograph of such a destroyed pattern is shown.

This limitation of lift-off methods is due to the resist characteristics. The use of a better resist with excellent temperature characteristics would solve this problem. An alternative method is the use of polyimide as a first layer. Fully cured polyimide withstands temperatures up to 400°C without deformation or solvent evaporation. However, to remove this polyimide, very strong solvents are necessary, so that the removal and lift-off step becomes critical. An alternative solution to the thermal instability of the resist is better cooling of the wafer during sputtering.

Because of the complexity, this lift-off method is only interesting for the patterning of a single metal pattern, when a very high resolution is needed. For sensor applications, a simple process is preferred. Nevertheless, this three-level lift-off process combines well with a bipolar process.

4.2.3   A CVD oxide assisted lift-off method

The lift-off method described above is complicated and not sufficiently compatible with CMOS techniques. For example, two masks are necessary to apply an extra metal layer on top of a CMOS wafer. One for the etching of the contact holes and one for the metal pattern. The method described in this section has been optimized for CMOS compatibility and simplicity. Only one mask step is necessary to etch the contact holes and to lift the metal.

The process flow of this CVD oxide assisted lift-off method is shown in figure 4.14. The following steps can be observed.

1. Substrate with polysilicon and CVD oxide

This lift-off method is designed as a CMOS-compatible process; so, there is a polysilicon and a CVD oxide layer on top of the wafer to start with. A typical thickness for the polysilicon is 0.5 μm; for the CVD oxide 1.3 μm.

2. Spinning, exposure and resist development

On top of this structure, AZ 1350 H resist is spun in a standard photolithographic step. This resist is then exposed and developed. The postbake of the resist is carried out at 120 °C to decrease the under-etching during the HF etch step and to make the lift-off structure more temperature resistant.

3. Wet etching of the CVD oxide.

The CVD oxide is etched in a buffered hydrogen fluoride (BHF) solution or a special CVD oxide etchant until the polysilicon is reached. This step corresponds to a normal contact hole etch. By the isotropic characteristics of the wet-etchant solutions, the CVD oxide is under-etched by approximately the same distance as the thickness of the CVD oxide layer. This results in a nice mushroom-like, reverse-taper lift-off profile.

4. Deposition of the metal.

The same resist layer that was used for the etching of the contact area is now used as lift-off layer. This means that there is only one mask and one photolithographic step necessary for the etching of the hole and the patterning of the metal. The metal can be deposited by evaporation or by sputtering. A low sputter power has to be used for sputtered layers to avoid a wafer temperature rise above 100 °C.

5. Resist removal and lift-off of metal.

The resist is removed in acetone. To speed up the lift-off, ultrasonic agitation is used. The CVD oxide is not removed, as this layer is also necessary for the insulation of the CMOS structures on the wafer.

In figure 4.15 the cross-sectional profile of the lift-off structure is shown after evaporation of 0.5 μm of aluminium. A mushroom-type profile is observed. Due to the high temperature pre-bake step, the resist layer has flowed, resulting in rounded resist edges. The under-etching of the CVD oxide is approximately 2 μm. This distance is determined by the over-etch time (typically 1 min). Due to this over-etching, a small ribbon of polysilicon is exposed; for some applications this ribbon can interfere. For electrochemical sensors, however, no interference of this polysilicon ribbon is seen (see section 5.2.1).

With this method a resolution of 5 μm has been achieved. In figure 4.16

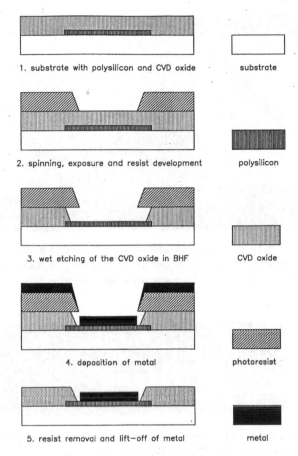

1. substrate with polysilicon and CVD oxide   substrate

2. spinning, exposure and resist development   polysilicon

3. wet etching of the CVD oxide in BHF   CVD oxide

4. deposition of metal   photoresist

5. resist removal and lift-off of metal   metal

Figure 4.14 The process flow of the CVD oxide assisted lift-off method.

a microphotograph is shown; the line-width/spacing is 10 μm. Smaller lines down to 5 μm can be produced with this technique; however, the reproducibility of the electrode area will be bad.

The following electrode materials have been patterned with the lift-off method described above: Ag, Ti/Ag, Cr/Ag, Ti/Pd/Ag, Ti/Pd/Ag/AgCl, Au, TiW/Au, Ti/Pd/Au, Pt, Ti/Pd/Pt, C, Al and Ti/Al. The only problems encountered with this technique are the thermal instability of the resist during sputtering, leading to destruction of the lift-off profile and to the formation of blisters. With some sputter parameters for platinum, a highly stressed metal layer was deposited. This stress, in combination with the thermal instability of the resist, resulted in the loss of the resist–CVD-oxide adhesion during sputtering.

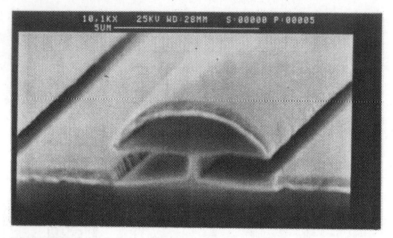

Figure 4.15   SEM photograph of the CVD oxide assisted lift-off profile after evaporation of 500 nm Al. Due to the post-bake of the resist at 120 °C, flowing of the resist occurred, resulting is a curved resist profile.

Figure 4.16   Microphotograph of a 500 nm thick evaporated gold test pattern (line-width/spacing= 10 μm/10 μm).

4.2.4 Conclusion

The thermal instability of the photoresist is a typical disadvantage of a lift-off process; the advantages, however, make lift-off the ideal patterning method for noble-metal electrodes in sensor applications. In the CMOS-compatible electrochemical sensor process described later in this book, the CVD oxide assisted lift-off method detailed above has been used with success for a great variety of metal layers. This flexibility is only possible with a lift-off method and cannot be achieved with a classical wet or dry etch method.

4.3   MICROMACHINING TECHNIQUES IN SILICON

Single-crystal silicon has interesting mechanical properties such as a Young's modulus comparable with stainless steel, a Knoop hardness close to quartz and a tensile yield strength three times higher than stainless steel wire (Petersen 1982). In contrast to most metals, single-crystal silicon yields catastrophically rather than deforming plastically. This brittle characteristic is enhanced by the single-crystalline nature of the material; due to small imperfections at the surface or in the bulk of the material, stress concentrations lead to cleavage of the silicon along crystallographic planes. So, although single-crystal silicon is a brittle material, as is known from wafer breakage, it is very rigid as a single die and extremely elastic. These properties make silicon an exciting mechanical material, especially when taken into account that there is an important group of interesting etchants for the fabrication of micromachined silicon structures on a mass production basis.

Doped silicon also has piezoresistive properties. A great variety of commercial pressure transducers and accelerometers are based on this piezoresistive effect. Nowadays several millions of pressure sensor dies and complete devices are shipped per year (NovaSensor 1987), [IC Sensors 1987]. This proves that the introduction of planar IC techniques on a mass-production basis can lead to commercial success in the field of silicon-based sensors.

Recent advances in permanently connecting two wafers to each other (Maszara 1991) have also opened the way to real three-dimensional multi-layer structures (Petersen *et al* 1991). Up to now this technique has been succesfully in the realization of mechanical sensors, but it can also be applied to chemical sensors.

Silicon, and especially silicon nitride, is inert to most chemicals and has a very high corrosion resistance. It is accepted that silicon and silicon nitride are inert and have biocompatible characteristics. Therefore, silicon is also an interesting material for the packaging of planar chemical sensors for industrial and biomedical applications. A new packaging method, based on micromachining techniques, will be presented in section 5.1.3.

From these initial remarks, one can conclude that there is an extensive need for the three-dimensional structuring of silicon for mechanical sensor development as well as for chemical sensor development. A good overview of the different etchants for silicon and an extended selection of micromechanical devices can be found in (Petersen 1982) (one of the standard works on micromachining in silicon). Other interesting papers on this topic are (Weirauch 1975, Kendall 1975, Kern 1978, Bean 1978, Seidel and Gsepregi 1983, Wu *et al* 1986 and Seidel 1987).

4.3.1   Isotropic etchants

Etchants for silicon are divided into two categories: isotropic and anisotropic etchants. The etch rate for different crystal directions is identical for isotropic etchants. Depending on the agitation of the etchant solution,

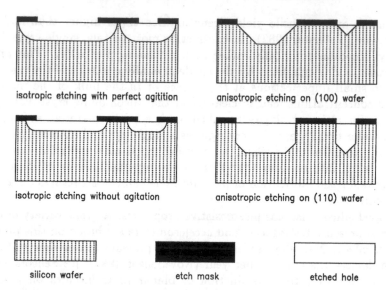

isotropic etching with perfect agitition

anisotropic etching on (100) wafer

isotropic etching without agitation

anisotropic etching on (110) wafer

silicon wafer

etch mask

etched hole

Figure 4.17 An overview of different etching profiles according to the anisotropic or isotropic nature of the etchant (adapted from Petersen (1982)).

a circular hole or a rounded box is etched into the silicon (figure 4.17). The under-etch of the etch mask is typical for isotropic etchants; this under-etch is a disadvantage for deep etching.

Anisotropic etchants are characterized by a different etch rate for the different crystal directions, resulting in structures with well defined surfaces, namely the crystal planes (figure 4.17).

The best known isotropic etchant for silicon is a mixture of HF, HNO_3, water and/or acetic acid. This etch system is very complex and etch characteristics vary dramatically with mix ratio, temperature, agitation and dopant concentration (Schwartz and Robbins 1976). Very high etch rates (> 50 μm min^{-1}) can be obtained; however, good control of the etch process is difficult. SiO_2 and photoresist can be used as an etch mask for relatively short etch times. For long etch periods, Au or Si_3N_4 can be used. The flatness of the bottom of the etched hole is generally poor, as the flatness is defined by agitation. Because of the isotropic nature of the etchant, the mask is under-etched so that these etchants can only be used satisfactorily for shallow etching (<10 μm).

The mixture (1250 ml HNO_3 65%, 500 ml H_2O, 37 ml BHF), as used for the wet etching of polysilicon, is a good mix ratio for a shallow etch in silicon. The etch rate for single-crystal silicon is approximately 0.1 μm min^{-1}.

4.3.2   Anisotropic etchants

Anisotropic (or crystal-orientation dependent) etchants are typically alkaline solutions at elevated temperatures (e.g. hydrazine, ethylenediamine or potassium hydroxide solutions in water); they have a different etch rate for the different crystal directions. Silicon has the same crystal structure as diamond, so the three important crystal directions are (100), (110) and (111). An anisotropy ratio that classifies the etchant can be defined as:

$$\text{anisotropy ratio (AR)} = \frac{\text{etch rate in } \langle 100 \rangle \text{ Si}}{\text{etch rate in } \langle 111 \rangle \text{ Si}}. \tag{4.1}$$

The anisotropy ratio equals 1 for isotropic etchants and can be as high as 400 for anisotropic etchants. Of course, the structure of the etched cavity depends on the wafer orientation, as depicted in figure 4.17. Since the $\langle 100 \rangle$ and the $\langle 110 \rangle$ planes etch much faster than the $\langle 111 \rangle$ planes, a pyramid structure is formed in (100) wafers, and boxes with very steep walls are formed in (110) wafers. The most interesting structures can be made in (100) wafers—standard wafers for CMOS processing—so that only the aspects of etching in (100) wafers will be discussed here.

As seen in figure 4.18, pyramids and cavities with thin bottom membranes can be etched in (100) wafers, depending on the etch time and the opening in the etch mask. For small openings a pyramid-shpaed hole is etched, bounded by the four $\langle 111 \rangle$ planes, as the anisotropic etch stops at $\langle 111 \rangle$ planes. For large openings, a cavity with $\langle 111 \rangle$ walls and a $\langle 100 \rangle$ bottom plate is etched. When the etch depth approaches the thickness of the wafer, a thin flexible silicon membrane is formed. As shown in figure 4.18, no under-etching of the etch mask is seen. Nevertheless, this is only true for concave mask openings with sides perfectly aligned to the (110) directions. This means that the patterns have to be aligned to the flat of the wafer, as this flat indicates the (110) direction. In figure 4.19 is shown what happens with different mask openings after a sufficiently long etch time. A square opening results in a pyramidal hole. Rectangular openings result in a V-groove. Patterns that are misaligned to the crystal directions result in a larger pyramidal pit. As a general rule, it can be stated that if the silicon is etched long enough, any closed shape in the mask will result in a rectangular pit in the silicon, bounded by the $\langle 111 \rangle$ surfaces and oriented in the (110) direction. The dimensions of the rectangle are such that the shape is perfectly inscribed in the resulting rectangle (Petersen 1982).

The best known anisotropic etchants are:

1. concentrated potassium hydroxide (KOH)
2. ethylenediamine pyrocatechol (EDP)
3. hydrazine (H_2N_4).

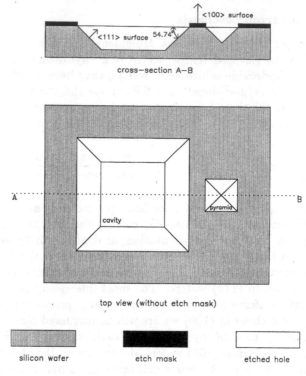

Figure 4.18   Cross section and top view of a (100) wafer etched in
an anisotropic etchant. Pyramidal holes and cavities are formed.

The important characteristics are summarized in table 4.1. For safety
reasons, etching with hydrazine should be avoided. The most interesting
etchant from table 4.1 is without doubt potassium hydroxide. This etchant
is not toxic, there is no explosion risk and the etch characteristics are excellent. The disadvantage of this etchant are the need for a Si_3N_4 etch
mask for long etch periods. The etch rate for thermal SiO_2 is 2 nm min^{-1}.
However, experiments have shown that a 1.5 μm thick wet oxide is not sufficient for the complete etching of a 380 μm thick wafer (6 hours) because
of pinholes in the oxide. The etch rate of LPCVD Si_3N_4 (a 120 nm thick
layer), etched for 6 hours, is not measurable. Another disadvantage of this
etchant is the presence of sodium and potassium ions in the solution. These
ions are known for their catastrophic influence on the mobile oxide charge
in MOS transistors, so they should be avoided in standard processing environments. Taking extreme care to avoid contamination and with extensive
cleaning of the wafers after the KOH etch, no danger exists.

    In figure 4.20, the influence of the temperature on etch rate and on
surface roughness is shown. All experiments were carried out in a reflux
system. The etch rate and the surface roughness were determined with a

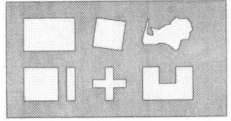

mask pattern before anisotropic etch

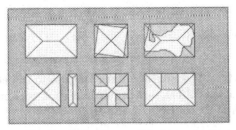

resulting structures after a sufficient long etch time

Figure 4.19   Resulting structures for different mask openings (anisotropic etchant in (100) Si).

Table 4.1   Principal characteristics of different etchants. (ED=ethylenediamine, P=pyrocathechol).

	KOH	EDP	H_2N_4
composition	7 M in H_2O	300 ml ED 80 gr P 100 ml H_2O	160 ml H_2N_4 40 ml H_2O
temperature [°C]	80	95 (115)	100
etch rate (μm/min)	1.2	0.8 (1.5)	1.5
AR value	400:1	40:1	
etch mask	Si_3N_4	SiO_2	SiO_2
surface finish	very smooth	rough	smooth
boron etch-stop	$> 10^{20}$ cm^{-3}	$> 7 \times 10^{19}$ cm^{-3}	no dependence
explosive	no	yes	extremely
toxic	no	yes	yes

Dektak surface profiler. The etch rate increases with increasing temperature and the surface roughness decreases with increasing temperature, so etching at high temperatures gives the best results. In practice an etch temperature of 80°C to 85°C is used to avoid solvent evaporation and temperature gradients in the solution.

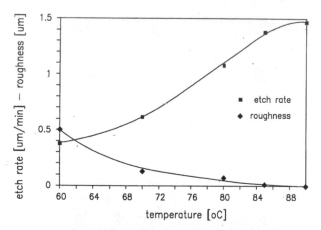

Figure 4.20   The influence of the temperature on etch rate and on surface roughness. The surface roughness has been measured for an etch depth of 60 μm (KOH 7M in H_2O)

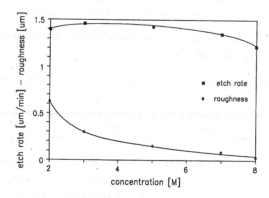

Figure 4.21   The influence of the KOH concentration on etch rate and on surface roughness (KOH in H_2O, temperature 85°C).

In figure 4.21, the influence of KOH concentration on etch rate and on surface roughness is shown. The etch rate is maximal around 4 M. The surface roughness, however, decreases with increasing concentration. Since the difference in etch rates for different KOH concentrations is small, a highly concentrated KOH solution (e.g. 7 M) is preferred to obtain a smooth surface. In figure 4.22, an SEM picture is shown of cavities etched in (100) silicon with KOH 7 M at 80°C.

To diminish the convex undercutting, 2-propanol is added to the solution. However, with this additive the formation of pyramids in the etched regions is observed during the first hours of the anisotropic etch, as shown in figure 4.23. These pyramids disappear after prolonged etching under continuous intense agitation. The addition of 2-propanol also diminishes

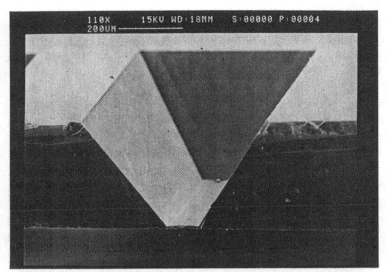

Figure 4.22 SEM photograph of a cavity etched in (100) silicon with KOH.

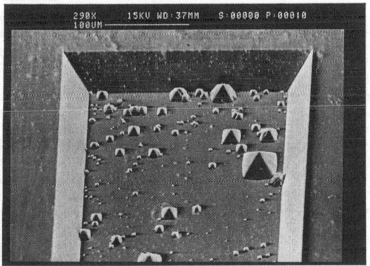

Figure 4.23 SEM photograph of cavities etched in (100) silicon with KOH–2-propanol. Pyramids are formed in the etched regions.

the anisotropy ratio so that the under-etching of the etch mask at the (110) planes increases.

This implies that 2-propanol is only of interest for the etching of convex structures, as needed for accelerometers, for example, under the condition that a good etch-stop is available to exclude the effect of pyramid growth. More details on convex undercutting and on the influence of etch composition can be found in (Puers and Sansen 1989).

Experiments with ethylenediamine pyrocatechol have demonstrated that this etchant has bad etch characteristics, especially if the etch temperature is kept below 100°C due to safety precautions. Advantage of this etchant are the low etch rate for SiO_2, so that SiO_2 can be used as an etch mask, and also the low K^+ and Na^+ ion content.

However, the etch rate is lower and the surface roughness is higher than for KOH. Deep etching of silicon may also result in the deposition of etchant residues on the etched surfaces (Wu *et al* 1986). Combined with the explosion risk and the toxicity, it can be concluded that the use of EDP as an anisotropic etchant should be avoided.

Recently, attention has been focused on anisotropic etchants which are compatible with CMOS processing and which are neither toxic nor harmful. Good results have been obtained with ammonium hydroxide–water mixtures (Schnakenberg *et al* 1990) and tetramethyl ammoniumhydroxyde–water mixtures (Tabata *et al* 1991, Schnakenberg *et al* 1991).

4.3.3   Etch-stop techniques

To obtain a uniform thickness of the silicon membrane and a reproducible sensor, etch-stop techniques are necessary. This is mainly due to the non-uniformity of the silicon wafer thickness. The non-uniformity or taper of double-polished wafers can be as high as 40 μm and as low as 2 μm for excellent quality wafers. The etch rate itself is relatively reproducible as the etch process is reaction-rate controlled. A good control of temperature and etchant concentration is nevertheless necessary. Stirring has little effect on etch rate.

This is illustrated in figure 4.24. In this experiment a double-polished wafer was etched in KOH until a membrane thickness of approximately 90 μm was obtained. In figure 4.24 a map of the wafer thickness, the membrane thickness and the etch depth is shown. Uniformity plots are also shown. The uniformity plot is obtained by subtracting the overall average value from the measured value. It is seen very clearly on the uniformity plots that the variations in membrane thickness are mainly due to wafer non-uniformity. The variation in etch depth is typically 1%, as can be derived from the uniformity of the etch depth (figure 4.24). This variation in etch depth, combined with the wafer taper of 2 μm in the best situation, still results in a membrane thickness variation of more than 5 μm for 20 μm membranes on 3 inch wafers (a 3 inch wafer has a typical thickness of 380 μm). This tolerance on membrane thickness is too high for most sensor applications. Therefore an easy etch-stop technique is necessary for the industrial manufacturing of micromachined devices.

The most widely known etch-stop is the boron stop, based on the reduced etch rate of heavy boron doped regions. For KOH, a boron concentration higher than 10^{20} cm^{-3} is needed to reduce the etch rate by about 20.

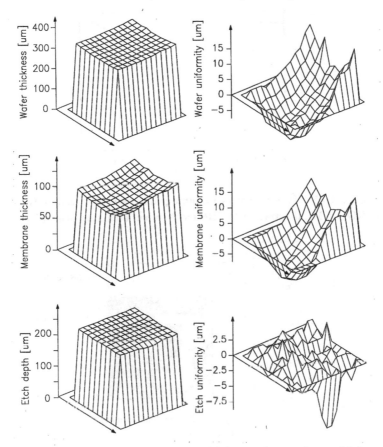

Figure 4.24 A map of the wafer thickness, the membrane thickness and the etch depth.

These extremely high boron concentrations are not compatible with standard CMOS or bipolar techniques, so they can only be used for structures without integrated electronics. The diffused layer is so heavily doped that no electrical devices can be made in this layer.

Electrochemical etch-stop techniques are based on the formation of a SiO_x passivated surface induced by an applied potential between the wafer and the etch solution. This phenomenon is known for KOH, EDP and H_2N_4 etchants and is covered extensively in the literature: e.g. (Jackson *et al* 1981, Palik *et al* 1985, Glembocki and Stahlbush 1985, Hirata *et al* 1985, Palik *et al* 1987 and Hirata *et al* 1987).

The etch-stop principle can easily be verified with electrochemical techniques. In figure 4.25 a linear-sweep voltammogram of an n-type silicon wafer in KOH 7 M is shown. This current–voltage relation was measured using the set-up depicted using figure 4.26. A complete 3 inch silicon wafer

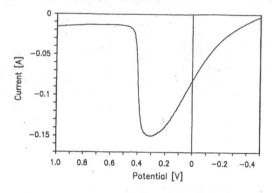

Figure 4.25   Linear-sweep voltammogram of n-type silicon (3 Ωcm in KOH 7 M), illustrating the electrochemical etch-stop principle.

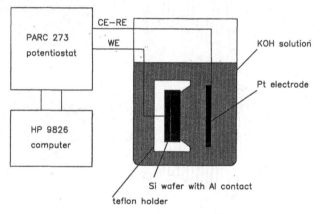

Figure 4.26   Measurement set-up for figure 4.25.

was exposed to the solution. A good ohmic aluminium contact was made on the backside of the wafer via an arsenic implanted region. This electrical contact was sealed from the etchant by a mechanical teflon holder and O-rings.

The voltage was applied to the silicon by a platinum electrode in a two-electrode configuration. It is better to use a three-electrode system with a calomel reference electrode, as the potential is not well defined with a two-electrode system. In our experiments however, a two-electrode system has been used for simplicity of the etch system. Measurements were carried out on the automatic PARC 273 potentiostat described in section 3.2.1. Three different regions can be determined in the voltammogram (figure 4.25).

In the first region the current increases as the potential increases. This can be explained by the formation of an anodic oxide on the surface of the silicon. The growth rate of this electrochemically induced oxide layer is proportional to the applied voltage and is smaller than the oxide etch rate

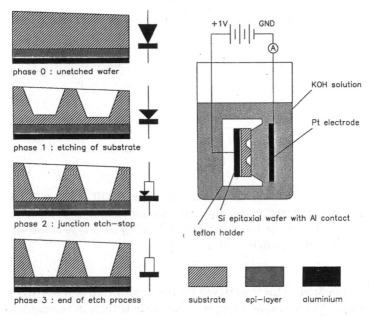

Figure 4.27   The electrochemical etch-stop technique applied to the formation of thin silicon membranes by the use of an epilayer p–n junction.

in the KOH solution, so the silicon is still etched away.

At a well defined potential (in this example 300 mV) the oxide growth rate equals the oxide etch rate; a further increase of the potential results in a steep fall of the current due to complete passivation of the silicon surface with the electrochemically induced oxide. At this potential range, etching of the silicon stops completely.

The third region (0.4 V–1 V) is characterized by a constant current as a function of the applied potential. Etching of the silicon is also completely stopped. The current in this region is necessary to compensate for the continuous etching of the passivating oxide. The thickness of this passivating oxide can be rated close to 1 nm, taking into account an etch rate of 1.4 nm min^{-1} for SiO_2 and a delay in the etching of the silicon of 30 to 60 seconds after the removal of the passivating potential. The same value, also derived from the time to etch back, has been reported for KOH etching at 60 °C by (Smith *et al* 1987) in a more elaborate study.

This principle, in combination with an epilayer, can be applied to the formation of silicon membranes with well defined thickness, even on wafers with an important taper (figure 4.27). An n-type epitaxial layer (phosphorus-doped, 10^{15} cm^{-3}) is grown on a p-type substrate (boron-doped, 30 Ωcm); this p–n junction on the complete wafer forms a large diode. The p-type substrate is then etched away and the etching is stopped

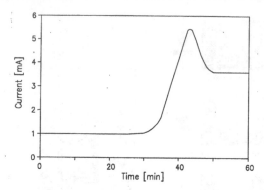

Figure 4.28   The current through the wafer/KOH interface is an excellent parameter to monitor the electrochemical p–n junction etch-stop process. (n-type epilayer (10^{15} cm^{-3}) on p-type substrate (30 Ωcm) etched in KOH 7 M at 80°C).

at the p–n junction. This is achieved by the application of a positive passivating voltage to the n-type epilayer. The diode is then in reverse bias and no current can flow, leaving the substrate at open circuit potential. The substrate is etched normally, as no passivation potential is present at the wafer/KOH interface.

However, at the specific moment that the p–n junction is reached, a current can flow and the applied potential passivates the epilayer so that the etching is stopped. Etching continues on the areas where the wafer is thicker until the p–n junction is reached there too. The thickness of the silicon membrane is thereby defined solely by the thickness of the epilayer. With this method, a uniformity better than 1 μm can be obtained on 10 μm thick membranes.

This process can be monitored very easily by measuring the current through the wafer/KOH interface. In figure 4.28 this current is plotted during the last hour of such an electrochemical etch-stop experiment. In this current-versus-time curve, three regions can also be distinguished.

In the first region, the substrate etches normally; this region corresponds to the situation depicted in figure 4.27 as phase 1. In this experiment a current of 1 mA still flows through the interface. This current is limited by the reverse bias current of the diode; as the area of this diode is as large as the complete wafer, this reverse bias current can be rather important, especially if there are defects at the epilayer/substrate interface. If the reverse bias current is too high, passivation and an unintentional etch-stop of the substrate occurs. This problem can be solved by the use of a fourth electrode (Kloeck and de Rooij 1987, Kloeck 1989, Kloeck *et al* 1989).

In the second region, corresponding with phase 2 in figure 4.27, etching stops at the areas where the p–n junction is reached, according to the principle described above. This region is characterized by first an increase in the current through the wafer/KOH interface due to charge transfer during the etch-stop, followed by a small decrease in current.

From the moment that all cavities are etched up to the p–n junction, the current remains constant. This current is due to the continuous etching of the passivating oxide. This third region corresponds to phase 3 in figure 4.27. The etch procedure can be stopped the moment that this current plateau is observed. From these observations it can be concluded that the measurement of the current through the wafer/KOH interface gives valuable information on the etch process and indicates when the procedure can be stopped.

This electrochemical p–n junction etch-stop method has been used with success for the manufacture of 15 μm thin silicon membranes. The electrochemical method has, in comparison with the boron etch-stop, the advantage that the dopant concentration of the epilayer is sufficiently low to be used in combination with CMOS or bipolar processing, so that electronic components can be made in the silicon membrane. The electrochemical p–n junction etch-stop has the disadvantage that the back of the wafer with the aluminium contact has to be sealed hermetically from the etchant solution. This has to be achieved with a mechanical fixture. Stress induced by this mechanical fixture and by the epilayer can result in breakage of some wafers during etching.

4.3.4   Conclusion

In this section an anisotropic etch process for micromachining in silicon is described using KOH as etchant and Si_3N_4 as etch mask. Ethylenediamine pyrocatechol and hydrazine has to be rejected due to the safety risks with these etchants. To control the etch process, electrochemical etch-stop techniques are a necessity.

Silicon micromachining is an exciting technique for the mass production of silicon-based sensors. Although mainly used for mechanical sensors until now, several groups are nowadays developing a silicon micromachined package for chemical sensors with these techniques, as will be discussed in section 5.1.3.

4.4   FORMATION OF PLANAR CHEMICAL MEMBRANES

The chemically active membrane determines the sensitivity and selectivity of electrochemical sensors. The quality of a sensor is thereby linked to the quality of the chemical membrane on top of the electrochemical detection unit. Planarization of chemically active membranes is therefore the most critical step in the development of planar or integrated chemical sensors

(Oesch and Simon 1987). Practically all chemical membranes are based on classical polymers, such as PVC, PVA, PHEMA, PVP, PUR, and on biological material, such as enzymes and proteins.

These materials can be applied on the wafer with various techniques; the most interesting are spin coating, dip coating, casting and the Langmuir–Blodgett film approach. The difficult part is the patterning of the resulting films and the adhesion of the chemical membrane to the substrate. No commercial techniques are known for the patterning of chemical membranes because of the very specific characteristics of these materials.

Good adhesion of the membrane to the silicon surface is also a problem. Mechanical anchoring based on a suspended polyimide mesh has been proposed by Janata *et al* as a long-term solution (Blackburn and Janata 1982). Recent research on adhesion improvement, however, is based on the surface pre-treatment with silanizing or covalent-coupling agents (adhesion promoters). These complex chemical compounds are spun onto the wafer as a monomolecular layer; a covalent bond is formed between the substrate and one side of the silanizing agent. One the other side, a covalent bond is formed with the planar membrane material. This covalent coupling guarantees an excellent long-term adhesion.

In this section, an overview of possible techniques for the mass production of planar chemical membranes for electrochemical sensors will be given; most of these techniques are derived from standard microelectronic procedures, adapted to the specific needs of chemical sensing. For biosensors, enzyme membranes are extremely important. Enzyme membranes and immobilization techniques are therefore subsequently discussed.

4.4.1   Enzyme membranes and immobilization techniques

Without an enzyme membrane or a membrane containing biological material, no reliable biosensor can be made. Therefore, the enzyme has to be immobilized in a membrane; the characteristics of this membrane are of critical importance for the functionality of the glucose sensor. For planar enzyme membranes, not only the enzymatic characteristics are crucial, but also adhesion to the detector structure and patterning capabilities are necessary. For planar sensors, adhesion has to be achieved with chemical binding techniques, since no mechanical fixtures, such as an O-ring, can be used.

Immobilization of an enzyme has two important advantages.

1. A mobile enzyme can leach out of the enzyme membrane. This will influence the response of the sensor and diminish the lifetime of the device. For 'in vivo' applications, leaching of the enzyme out of the sensor cannot be tolerated since small amounts of glucose oxidase might interfere in the metabolism.

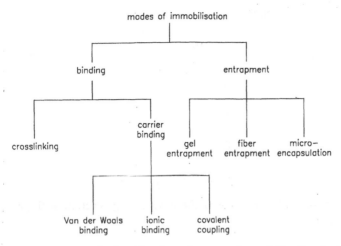

Figure 4.29 A classification of enzyme immobilization techniques.

2. By immobilization, the stability of the enzyme is improved, resulting in a longer lifetime of the device as the activity of the immobilized enzyme is retained longer. The shelf-life of the sensor is also increased.

Several techniques are known for the immobilization of enzymes; see, for example, (Woodward 1985) or (Guilbault *et al* 1991). In figure 4.29, a classification of immobilization methods is shown. A schematic representation of the different immobilization methods is depicted in figure 4.30.

One can distinguish between the following.

1. Crosslinking

A water-insoluble enzyme layer can be obtained by intermolecular crosslinking of the enzyme molecules in the absence of a carrier. Therefore, a bifunctional reagent such as glutaraldehyde is used. It is important that only functional groups on the enzyme that do not participate in the catalytical reaction are bound. In figure 4.31, the principle of crosslinking with glutaraldehyde is shown. This crosslinking is irreversible and is resistant to extreme temperatures and pH values.

Since enzymes are expensive, and in order to avoid high enzyme concentrations, inert proteins are added. These proteins, such as bovine serum albumin, are co-polymerized with the enzyme. The volume of the enzyme preparation can thereby be increased to the desired value, without great cost.

2. Van der Waals binding

Adsorption of enzymes onto a carrier based on Van der Waals forces is without doubt the most elementary method. This adsorption is realized by adding an enzyme solution to the carrier, followed by a wash step to remove the non-adsorbed enzyme. This binding, however, is reversible

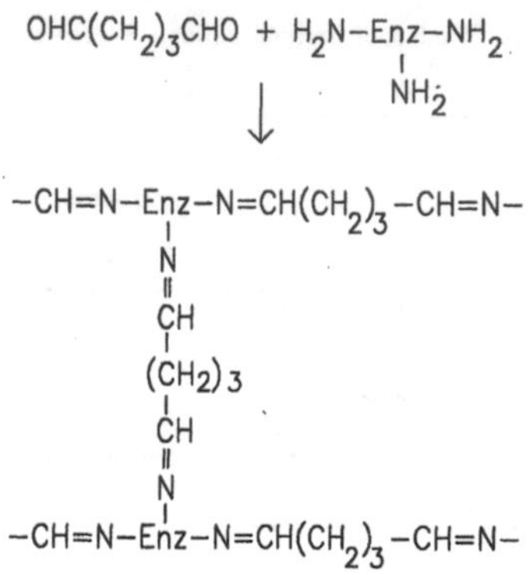

$$OHC(CH_2)_3CHO + H_2N-Enz-NH_2$$

$$\downarrow \quad NH_2$$

$$-CH=N-Enz-N=CH(CH_2)_3-CH=N-$$
$$| $$
$$N$$
$$\|$$
$$CH$$
$$|$$
$$(CH_2)_3$$
$$|$$
$$CH$$
$$\|$$
$$N$$
$$|$$
$$-CH=N-Enz-N=CH(CH_2)_3-CH=N-$$

Figure 4.30   The principle of the crosslinking reaction with glutaraldehyde.

and non-specific and is influenced by a large variety of parameters. The reproducibility of this method is therefore questionable.

3. Ionic binding

A carrier with residues of ion-exchange material is required for this method. An ionic binding is formed between the carrier and the enzyme under well defined conditions. The binding is stronger than the Van der Waals binding, but still reversible.

4. Covalent coupling

A rigid bond is formed by covalently coupling the enzyme to the carrier material. It is important that only functional groups on the enzyme that do not participate in the catalytical reaction are bound. Several coupling reactions are used in practice. Also, multifunctional reagents such as glutaraldehyde can be used for this purpose. Since a great variety of enzymes and materials can be coupled, this method is without doubt the most versatile carrier-binding method.

5. Gel entrapment

The enzyme molecule is entrapped in a gel matrix. The gel material can be, for example, agar, cellulose, collagen, PVA, PVP or polyacrylamide. To prevent leaching of the enzyme out of the gel matrix, this method can be combined with a chemical crosslinking step.

6. Fibre entrapment

Enzymes can also be entrapped in hollow fibres with a semi-permeable wall. Cellulose acetate fibres are used for this purpose in industrial processes.

7. Micro-encapsulation

Droplets of enzyme solution are encapsulated in semi-permeable microcapsules made of organic polymers. The substrate and the reaction products have to diffuse through the wall of the microcapsules.

Not all these techniques can be applied to the immobilization of enzymes

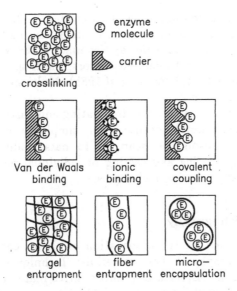

Figure 4.31   Schematic representation of the different enzyme immobilization techniques.

in planar membranes. The best-suited techniques are crosslinking of the enzyme with bovine serum albumin, gel entrapment and the direct covalent coupling of the enzyme to the electrode surface. Different possibilities are available for the deposition of these membranes. In the following sections, these techniques are reviewed and examples of typical applications are given.

4.4.2   Casting and dip coating

The simplest way to apply a membrane on a sensor is dip coating. A mounted sensor is dipped into a solution containing the polymer (the membrane material) and a solvent. After evaporation of the solvent, a thin membrane is formed on the surface of the sensor. This method results in membranes with poor uniformity and reproducibility. This technique is interesting for prototyping, as the membrane composition can be changed very easily without the need for a complete wafer for each experiment.

A similar method is known as casting. This technique is used, for example, for the preparation of emulsions on photographic films. Casting is based on the application of a given amount of dissolved material on the surface of the mounted sensor or wafer using, for example, a precise micro-pipette. This method gives a more uniform and a more reproducible membrane than dip coating.

Both methods have important disadvantages. As a mass-production technique, dip coating is not appropriate due to the poor uniformity and reproducibility. Dip coating, as well as casting, suffers from the lack of patterning capabilities, so these methods are not suited for wafer-scale use. By casting small volumes in little cavities made in dry resist or thick poly-imide layers on individual mounted sensors, multi-species sensors can be made (Sibbald *et al* 1985, Kimura *et al* 1985). Multi-species sensors cannot be made with dip coating.

In spite of these disadvantages, dip coating and casting are widely used in the planar sensor world because of the simplicity of the method. The claim that planar sensors are cheap due to mass-production techniques is then false, as dip coating and casting of individual mounted sensors are expensive manual processing techniques.

4.4.3   Photolithography

An elegant approach for the preparation of planar chemical membranes is the photolithography of these materials. Techniques such as spinning, UV exposure and development of photosensitive materials are well known, low-cost, mass-production procedures. The inherent patterning capability makes photolithography the ideal membrane technology.

The simplest way is the inactivation of specified regions in a spin-coated membrane by UV exposure. This method is described by (Kuriyama *et al* 1986), for example, for the fabrication of a urease sensor using a differential enzyme FET set up.

Another obvious methodology is the incorporation of chemically active substances in commercially available resists, as detailed by (Lauks 1981) for example. In his work he experimented with the addition of ionophoric groups, such as valinomycin, to standard Kodak KMER photoresist to form ion-selective membranes. A disadvantage of this approach is the impurity of the commercial resist. All kinds of unknown substances are added to improve the microelectronic characteristics of commercial resist layers. These substances influence the chemical behaviour of the membrane in a random way .

To cope with this problem, one has to prepare the photosensitive material itself from high-purity basis material. This method has been used with success by different authors (Hanazato *et al* 1986, Engels and Kuypers 1983, Moriizumi and Miyahara 1985, Ichimura and Watanabe 1982). As a basis material, water-soluble polymers such as PVA, PVP and PHEMA are used. These polymers are dissolved in water and a photosensitizer is added. After spin coating, the polymer film is photochemically crosslinked by UV light. These photosensitive systems act as a negative resist. In the exposed regions, an insoluble hydrogel is formed. The development or removal of unexposed regions is carried out in (warm) water. A well known system is PVA in combination with $(NH_4)_2Cr_2O_7$ as a photosensitizer (Grimm *et al* 1983). Nevertheless, this system is not ideal for enzyme membrane prepa-

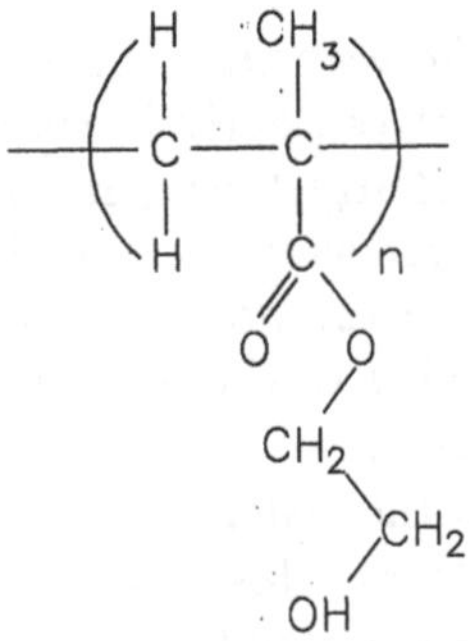

Figure 4.32   The structure of polyhydroxyethylmethacrylate.

ration due to the oxidizing effect of the photosensitizer so that it cannot be used for glucose oxidase membranes. A better approach is the use of stilbazolium groups as a photosensitizer (Ichimura and Watanabe 1982, Moriizumi and Miyahara 1985). However, photosensitizers with stilbazolium groups are not commercially available, limiting the use of this system. Another Japanese group claims that the use of a PVP membrane that is easy to photocrosslink is a better technique for enzyme sensor membranes (Hanazato *et al* 1986).

In the sensor research at K U Leuven, PHEMA is studied as a candidate for a photosensitive matrix. The polymer PHEMA (van Hooghten 1987) can be used as a hydrogel layer for oxygen sensors and ion-selective electrodes and as a glucose oxidase membrane for glucose sensors. It has the advantage that this material will rehydrate more easily in comparison with crosslinked PVA. The rehydration of crosslinked PVA is difficult because of the crystalline nature of this material. It is also used for biomedical applications; for example, soft contact lenses are made in crosslinked PHEMA. In order to obtain good biocompatibility of these contact lenses, they have to be hydrophylic and permeable to oxygen. Similar requirements are necessary for planar hydrogel membranes. The structure of PHEMA is shown in figure 4.32.

The study of the photolithography of PHEMA hydrogels is carried out in cooperation with Professor H Berghmans (head of the laboratory for polymer research, Faculty of Science, K U Leuven). For this purpose a linear PHEMA has been synthesized. A linear PHEMA is preferred over polymerization of the HEMA monomer as the hydrogel material has to be spin coated, crosslinked in situ and patterned. A premature crosslinking of the HEMA monomer in an unordered way has to be prevented in order to allow photolithography of the hydrogel.

For the photolithography of PHEMA, the following procedure is used (a similar process can be used for PVA). The process can be summarized as follows.

1. Preparation of the photosensitive PHEMA solution

Since linear PHEMA dissolves poorly in water, methanol is used as a solvent. The photosensitizer $(NH_4)_2Cr_2O_7$ is firstly dissolved in methanol. A saturated solution is preferred as the maximal solubility of $(NH_4)_2Cr_2O_7$ in methanol is low (0.2%). For the preparation of the photosensitive PHEMA mixture, 10% PHEMA is added to this saturated solution.

2. Spin coating of the wafer

This mixture is spun on the wafer at 1000 rpm for 30 seconds, resulting in 0.8 μm thick layers. If thicker layers are required, multiple coatings can be spun. In order to prevent striations, it is recommended to filter the mixture through a 1 μm filter prior to use.

3. UV exposure

The wafer with the PHEMA layer is dried at room temperature. Exposure is carried out in a normal mask aligner. A typical exposure time is 60 seconds if a 5 mW cm^{-2} UV source with a 254 nm wavelength is used. In the exposed regions, the PHEMA layer is crosslinked and becomes insoluble in methanol.

4. Development in methanol

The wafer is developed in methanol. During this process, the unexposed PHEMA is dissolved and removed.

This process produces excellent planar PHEMA membranes. In figure 4.33, a microphotograph of a patterned PHEMA layer is shown. A resolution better than 5 μm has been achieved on 1 μm thick layers. The adhesion of the PHEMA hydrogel to the SiO_2 surface is also excellent because of the crosslinking of the PHEMA with surface hydroxyl groups. This reaction, however, requires a sufficient exposure time. The result of an inadequate exposure time is shown in figure 4.34. Although the PHEMA is crosslinked and pattern formation is observed, adhesion is insufficient to hold the membrane. It appears that the PHEMA layer only adheres at the edges of the pattern. Figure 4.33 shows a micro-photograph of a patterned PHEMA layer.

Since in this example an oxidizing photosensitizer is used, this process cannot be used for gel-entrapped enzyme layers. Further improvement is therefore necessary. This includes the incorporation of photosensitive groups and enzymes into the linear PHEMA structure and characterization of the rehydration and water-retention process. The main problem with these photosensitizers, such as those described by Ichimura and Watanabe (1982) and Moriizumi and Miyahara (1985), is their limited commercial availability up to now. Photolithography of bio-materials, however, is becoming more and more popular (Schoen *et al* 1990, PhotoLink 1991), hence major progress is expected in this field.

The use of a photolithography-based photosensitive polymer system is the most elegant approach for the preparation of planar chemical membranes. It is also the most difficult one. A lot of basic and advanced chem-

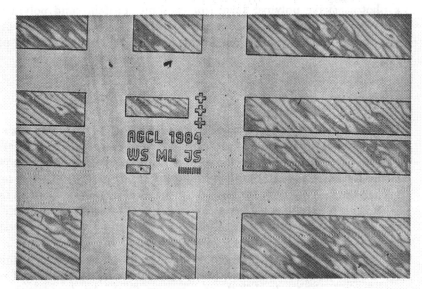

Figure 4.33   A microphotograph of a patterned PHEMA layer.

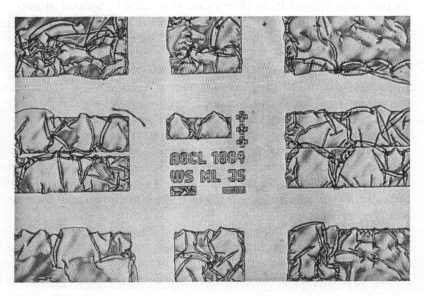

Figure 4.34   The result of an inadequate exposure time on the patterning process of PHEMA.

istry is needed for the preparation of and the research into photosensitive planar chemical membranes.

Not all the different chemically active membranes can be converted into a photosensitive polymer layer (for example, proteins and polymers like PVC and Teflon cannot be converted). Other microelectronics techniques have to be adapted for this purpose; good examples are lift-off and plasma etching.

4.4.4   Lift-off

Lift-off has been used by (Kuriyama *et al* 1986) for the pattening of a bovine serum albumin matrix based glucose oxidase membrane. A modification of this process is discussed in this section. A problem with lift-off, however, is the thickness needed for the photoresist layer. A typical resist layer is 1.3 μm thick. The thickness needed for chemical membranes can be as high as 50 μm. Therefore, thicker resist layers are necessary to pattern these thick layers. Also, lift-off can only be used with materials that are resistant to the solvent necessary for the resist removal.

A planar IC-compatible technique, based on lift-off of the BSA membrane, has been developed by the authors. Since the enzyme content of such a membrane is relatively low, the process was firstly developed without the GOD enzyme in the membrane to reduce experimental costs. The lift-off method was derived from (Kuriyama *et al* 1986). Special attention has been paid to adhesion of the membrane to the silicon surface, therefore a special pre-treatment of the sensor surface based on silanizing agents has been developed.

The crosslinking mechanism of bovine serum albumin and glutaraldehyde is similar to the mechanism shown in figure 4.31. After addition of glutaraldehyde to a bovine serum albumin solution, a gel is formed. The viscosity and the formation time of the gel are the most important characteristics for the preparation of membranes. They are mainly influenced by bovine serum albumin and glutaraldehyde concentrations and by temperature. Room-temperature processing combined with fast gel formation ($\pm$10 min) is preferred for the preparation of planar membranes. Low temperatures (4° C) and long gel formation times are sometimes used for very sensitive enzymes.

The influence of concentration on the gel formation time is shown in figure 4.35 and figure 4.36. At low protein and glutaraldehyde concentrations, no gel formation is observed. At high concentrations, a gel is formed immediately. So, by proper choice of concentrations, the desired gel formation time can be obtained. The viscosity of the solution is mainly determined by the protein concentration.

The following lift-off process for the deposition and patterning of bovine serum albumin based GOD membranes has been derived (figure 4.37).

1. Degreasing of the substrate in acetone
Acetone is used for wafer cleaning. The standard cleaning procedure in

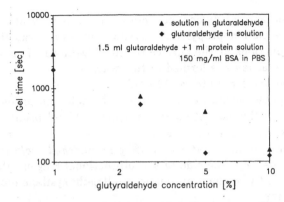

Figure 4.35   The influence of glutaraldehyde concentrations on gel formation time of BSA layers.

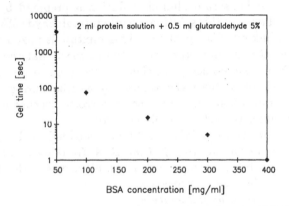

Figure 4.36   The influence of bovine serum albumin (BSA) concentrations on gel formation time of BSA layers.

fuming HNO_3 cannot be followed as this would attack the underlying metal layers of the electrochemical electrode structures.

2. Photoresist spinning, UV exposure and development

Photoresist is applied to the wafer according to standard procedures. This results in a 1.4 μm thick resist layer. It allows lift-off of bovine serum albumin membranes with a thickness up to 1 μm. For thicker membranes, thicker resist layers are required. This can be achieved by decreasing the rpm or by choosing another resist composition. It is important that the openings in the resist are completely free of resist residues. Therefore, a plasma cleaning may be required (100 W, O_2, 60 s).

3. Spinning of adhesion promoter

As adhesion promoter 1% γ-amininopropyltriethoxysilane in 1,1,2,trichlorotrifluoroethane is used. This solution is spun onto the wafer at 5000 rpm for 30 seconds. The normal solvents for this silanizing agent, such as methanol or xylene, can not be used as they attack the photoresist layer. γ-amininopropyltriethoxysilane forms a covalent bond with the hydroxyl functions at the surface of the SiO_2 layer. It is assumed that a monomolecular layer is formed in this way.

4. Baking for 5 minutes at 110 °C

This step was introduced to activate the adhesion promoter. The exact mechanism of this activation is not known. Good results have also been obtained without this baking step.

5. Soaking for 15 minutes in a 5% glutaraldehyde solution

During this step, one side of the bifunctional reagent glutaraldehyde is linked covalently to the γ-amininopropyltriethoxysilane molecules.

6. Washing

The wash step removes all excess of glutaraldehyde.

7. Spinning and crosslinking of the BSA membrane

A typical bovine serum albumin solution is prepared by dissolving 3 g BSA in 10 ml PBS. To this solution 0.5 ml glutaraldehyde 5% in PBS is added for the crosslinking of the proteins. This mixture is mixed and cast on the wafer. The gel formation starts from the moment that the glutaraldehyde is added. From the moment that this gel formation results in a perceptible increase in viscosity, the wafer is spun at 50 rpm during 30 s. With the indicated composition, this increase in viscosity occurs after 8 minutes. After 20 minutes the gel is completely formed.

The same procedure is followed if GOD is added to the membrane.

8. Resist removal and lift-off of the BSA–GOD membrane

The resist is dissolved in acetone in an ultrasonic bath. The BSA membrane is lifted during this process .

9. Lyophilisation, or freeze-drying

If the sensor is not used immediately, the membrane is lyophilized, or freeze-dried. The sensors are stored in a refrigerator at 4 °C after lyophilization. This will increase the storage time and prevent decay of enzyme activity.

A resolution of 10 μm can be achieved with this method; this is adequate for enzyme membranes. To obtain a higher resolution, a lift-off profile with overhang is necessary. In figure 4.38 a microphotograph of a lifted BSA layer is shown. The local uniformity of the membrane is not perfect; striations in the layer are observed. This is mainly related the low rpm and particles in the enzyme solution. If desired, the uniformity can be further improved by proper choice of viscosity and by filtration of the enzyme solution. These membranes with GOD added have been used with success for the realization of a planar glucose sensor (see section 5.3.2).

4.4.5 Plasma etching

Plasma etching is a widely used patterning method, based on the combined

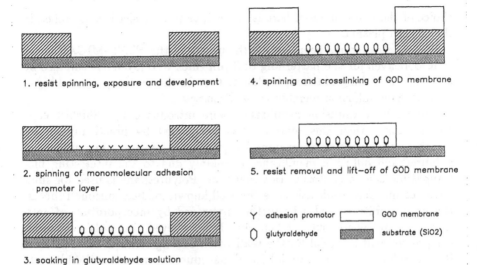

1. resist spinning, exposure and development

2. spinning of monomolecular adhesion
   promoter layer

3. soaking in glutyraldehyde solution

4. spinning and crosslinking of GOD membrane

5. resist removal and lift—off of GOD membrane

Y adhesion promotor ☐ GOD membrane

○ glutyraldehyde ▨ substrate (SiO2)

Figure 4.37   The different process steps for the lift-off technique applicable to bovine serum albumin membranes.

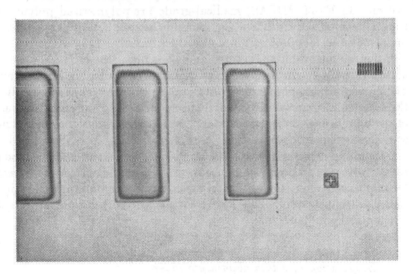

Figure 4.38   A microphotograph of a lifted bovine serum albumin layer.

action of bombarding ions and reactive species in a gas plasma. Selectivity and etch rate are mainly determined by the gas composition. Some examples are CF_4 and C_2F_6 for SiO_2 or Si_3N_4, and CCl_4 for Al. For organic material and polymers an oxygen plasma is used; the material is first oxidized, then converted into volatile components and pumped away. The choice of the mask material here is critical, as photoresist is also etched in the oxygen plasma.

Plasma etching has been used with success in the ESAT–MICAS group of K U Leuven for the patterning of 3 μm thick PVC layers. It can also be used for the etching of teflon, Nafion or PUR. In this section, dry-etching of polyurethane diffusion membranes is discussed.

Polyurethane diffusion membranes were introduced by (Shichiri *et al* 1982). Since then this material has been used by practically all research groups working on enzyme-based glucose sensors. The success of polyurethane diffusion membranes is mainly due to the interesting biocompatible characteristics of this material. Polyurethane is used for similar biocompatible applications as the well known medical silicone rubbers. Biocompatibility can also be further improved by incorporation of anti-thrombogenic agents such as heparin. It is believed that polyurethane is the pre-eminent material for the realization of long-term 'in vivo' sensors. It is therefore the ideal material for planar glucose sensors, as the ultimate goal for these sensors is also long-term 'in vivo' use.

Two different polyurethane materials can be distinguished: prepolymerized polyurethane rubbers and polyurethanes for in situ polymerization. In ESAT–MICAS, medical-grade pre-polymerized polyurethane from BF Goodrich Chemical Co, Belgium is used. This polyurethane is dissolved in tetrahydrofuran in a 4%-to-8% ratio.

If this polymer solution is coated on a glass plate, the resulting membrane is impermeable to glucose. When coated on top of a bovine serum albumin membrane, a microporous membrane is formed during the evaporation of the tetrahydrofuran solvent (Churchouse *et al* 1986). The exact mechanism of the pore formation is, however, not well understood and therefore needs some further investigation.

Polyurethane membranes can be applied by spin coating and can be patterned with dry-etching techniques in an oxygen plasma. As a patterning technique, a modification of the tri-level lift-off technique described in section 4.2.2 is proposed. This process involves the following steps shown schematically in figure 4.39.

1. Spin coating of the PUR membrane
2. Evaporation of a thin aluminium layer
A thin layer (100 nm) of Al is evaporated onto the PUR membrane. This layer will serve as a plasma etch mask.
3. Spin coating, exposure and development of the photoresist layer
A standard photoresist layer is spun over the Al layer. After exposure, the resist is removed in the illuminated regions.
4. Wet etching of the thin aluminium layer
The thin aluminium layer can be etched with a normal wet etchant. Since the Al layer is very thin and standard resist developer also etches Al, this

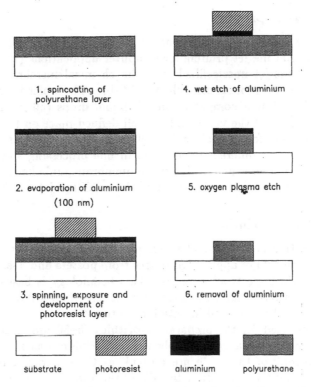

1. spincoating of
polyurethane layer

4. wet etch of aluminium

2. evaporation of aluminium
(100 nm)

5. oxygen plasma etch

3. spinning, exposure and
development of
photoresist layer

6. removal of aluminium

substrate   photoresist   aluminium   polyurethane

Figure 4.39   The different process steps for the dry etching of
polyurethane membranes

layer is also etched by a prolonged immersion in the developer solution.

5. Dry etching of the PUR *membrane and resist removal*

During dry etching in an oxygen plasma, the resist layer is removed and
the polyurethane is etched. From the moment that all resist is removed,
the Al layer serves as etch mask. Very thick layers of polyurethane can be
etched, since the resistance of Al to an oxygen plasma is excellent.

6. Removal of the thin aluminium layer

The aluminium etch mask is removed by immersion in the standard resist
developer. Developer does not attack the polyurethane layer.

With this process, planar PUR membranes have been realized with a
resolution better than 5 μm. It is therefore believed that dry etching of
polyurethane diffusion membranes in combination with lift-off of glucose
oxidase, bovine serum albumin membranes will result soon in a working
glucose sensor produced entirely with IC-compatible microelectronic-based
techniques. The use of these mass-production techniques for all layers of
a planar glucose sensor is an essential requirement for the realization of a
cheap, disposable sensor.

4.4.6   Ink-jet printing

This rather special technique is based on the same principle as that used for commercial ink-jet printers for computer applications (Kuriyama *et al* 1986). An ink-jet nozzle filled with the chemical membrane solution is placed above a computer-controlled XY table. The wafer is placed on top of this table. Under computer control, liquid drops (50 μm in diameter) are expelled out of the nozzle onto a well defined place on the wafer. This method is ideal for the fabrication of multi-species sensors, as different nozzles can print different membranes in one processing step. Problems encountered are poor uniformity of the membrane, unbalanced viscosity of the membrane solution and clotting of the small ink-jet nozzles.

4.4.7   Screen printing

Screen printing is a universal technique for the selective coating of flat surfaces. The field of application ranges from posters and T-shirts to high-precision hybrid electronic circuits. This technology is based on the application of a paste through openings in a screen on a flat surface. This technique will be discussed in detail in section 4.5.

It can be used for the preparation of thick (5–50 μm) membranes. In this case the membranes are applied as a thick-film paste. The viscosity and thixotropy of the paste are critical, as these parameters define the screen-printing characteristics. In spite of the appealing possibilities of this technique, very few research groups have published results on the use of screen printing for the preparation of planar chemical membranes (Pace *et al* 1985, Pace 1986).

At K U Leuven, research has been carried out by the authors in cooperation with Eurogenetics (Belgium) on the development of dipstick-type immunosensors with screen-printed layers. Therefore, a screen-printing process for the realization of porous layers has been developed. The paste consists of spherical particles in a hydrogel matrix; water is used as a solvent for these pastes. In figure 4.40, an SEM photograph of the realized structure is shown. Immunological components can be attached to the spherical particles or incorporated into the hydrogel matrix. Preliminary results indicated that these screen-printed layers can be used in dipstick-like immunosensors.

4.4.8   Conclusion

In table 4.2, some important properties of the different methods for the preparation of planar chemical membranes are summarized.

The most interesting method is without doubt photolithography. It is a low-cost process in the microelectronic environment, as all the equipment is already available. However, the chemistry is very complex so that up to

Table 4.2 Deposition and patterning techniques for planar chemical membranes.

method	typical use	thickness range (μm)	cost	uniformity	reproduci- bility	patterning
dip coating	universal	0.1 - 50	low	poor	poor	no
casting	universal	0.1 - 50	low	moderate	moderate	no
photolithography	pva, phema	1 - 10	moderate	good	good	yes
lift-off	proteins	0.1 - 3	moderate	moderate	good	yes
plasma etching	pvc, teflon	1 - 10	high	good	good	yes
ink -jet printing	universal	1 - 5	high	poor	moderate	yes
screen printing	universal	5 - 50	moderate	moderate	moderate	yes

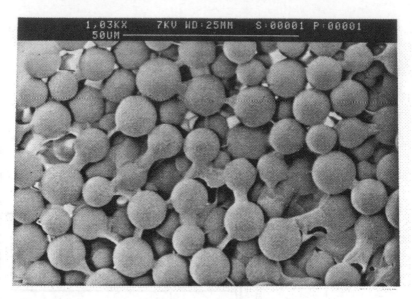

Figure 4.40   SEM photograph of a screen-printed porous layer suitable for the realization of dipstick-like immunosensors.

now only a few results have been published. Most of the few research groups (mainly Japanese) working on photolithography use proprietary photosensitizers which are not available for general use.

Plasma etching and lift-off are both a direct adaptation of basic microelectronic techniques. These methods can be applied for those membrane materials that cannot be patterned with photolithography.

Thick-film technology is ideal for the preparation of thick (> 10 μm) membranes. It deserves more attention since screen printing is a low-cost process with high flexibility in membrane composition, as long as the appropriate paste viscosity and thixotropy is obtained.

It can be concluded, that although several promising microelectronic techniques are available for the patterning of chemical membranes, most work on membranes for planar chemical sensors is based on casting and dip coating. These methods are cheap, easy and give good results on individual sensors. They are not suited for mass production due to poor uniformity and reproducibility. Lack of patterning capabilities exclude these techniques from use in the field of multi-species sensors.

4.5   THICK-FILM TECHNOLOGY

Thick-film or hybrid technology is based on the principle of screen printing. A paste with the appropriate viscosity and thixotropy is pressed onto the

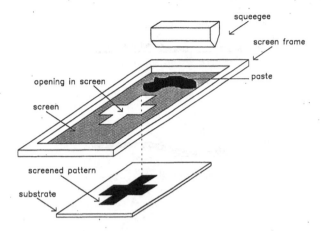

Figure 4.41 Schematic representation of the screen-printing process (adapted from Rikoski (1973)).

substrate with a squeegee through openings in the emulsion on a stainless steel screen (figure 4.41). The paste is a mixture of the material particles to be applied, an organic vehicle for adjusting the thixotropy and a solvent. The pattern made by photolithography in the emulsion on the screen is transferred onto the substrate in three steps (figure 4.42). On the first cross section, the paste is applied onto the screen and a substrate is placed under the openings in the emulsion. During the stroke, the squeegee lowers, pushes the screen onto the substrate and forces the paste through the openings in the screen onto the substrate during the horizontal motion. During the last step, the screen snaps back, the thick-film paste, which adheres between the screening frame and the substrate, shears and the printed pattern is formed on the substrate.

The resolution of this process is determined by the characteristics of the paste and the dimension of the openings in the screen. With a 325-mesh screen (this corresponds to a screen with 325 wires per inch, resulting in 40 μm openings in the screen) and a paste with optimum characteristics, a resolution of 100 μm can be obtained. For pastes that are difficult to print with a wire screen, a thin metal plate with openings can be used. The resolution of this method is, however, inferior (> 500 μm).

After printing, the wet films are allowed to settle for 15 minutes to flatten the surface and are dried. This removes the solvents from the paste. Subsequently the paste is fired in a conveyor furnace with a well defined temperature profile. During firing, the organic vehicle is burned off, the metallic particles are reduced or oxidized and the glass particles in the paste are sintered. The peak firing temperature ranges from 500 °C for overglaze to 1000 °C for conductors. After firing, the thickness of the film ranges from 10 μm to 50 μm.

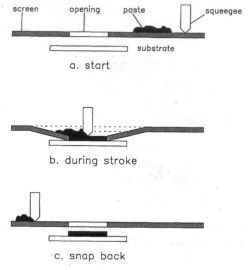

Figure 4.42   The three different steps during the screen-printing cycle (adapted from (Rikoski 1973)).

Different pastes are available commercially; namely conductive, resistive, overglaze and dielectric pastes.

The conductive pastes are based on metal particles, such as Ag, Pd, Au, Pt, or a mixture of these combined with glass. This glass is necessary for the adhesion of the metal conductor to the Al_2O_3 substrate. According to the bonding mechanism, three different categories are known.

1. Glass bonded or fritted pastes

Adhesion of the metal conductor is achieved with the addition of a glass mixture (30%). A typical composition of this fritted glass is 65% PbO, 25% SiO_2 and 10% Bi_2O_3.

2. Oxide or fritless-bonded pastes

Adhesion of the metal conductor is achieved with the addition of copper oxide (3%). This results in a film with a higher conductivity.

3. Mixed bonded pastes

Adhesion of the metal conductor is achieved with the addition of both a glass mixture and copper oxide.

As will be discussed in section 6.1, the composition of the paste is very important when used as an electrochemical electrode.

Resistive pastes are based on RuO_2 or $Bi_2Ru_2O_7$ mixed with glass (65% PbO, 25% SiO_2, 10% Bi_2O_3); the resistivity is determined by the mixing ratio. Overglaze and dielectric pastes are also based on a glass mixture; according to composition, different melting temperatures are obtained.

The technique described above can be designated as the conventional ceramic technology. Nowadays there also exist alternative technologies which can be well suited to sensor applications (Prudenziata 1991). Metal substrates coated with enamel are combined with special low-temperature firing pastes (500–650 °C). The major advantage of this technique is that the metal substrate can be machined in any form with normal equipment prior to the enamel coating. For sensor applications, thick-film technology based on polymer films is extremely important. Commercial devices, such as the ExacTech glucose sensor, are based on this technique. Special grades of polymer pastes (carbon, Ag and Ag/AgCl) are already available for the realization of (bio)sensors (Acheson 1991).

Other new development are based on the use of the green-tape technology. In this technology, special pastes are printed on non-fired, flexible ceramic material named 'green tape'. This tape is machinable with normal soft tools so that alignment holes, interconnection holes or cavities can be produced easily. Various layers can be punched and screen-printed separately. They then are laminated with pressure and heat to form a monolithic device. An example of a biosensor made with this quite novel technology is demonstrated in (Ruger *et al* 1991).

It is important to notice that for the different conductive, resistive or insulating pastes, the same screen printer and conveyor furnace are used. This implies that the investment necessary for the preparation of thick-film devices is low in comparison with semiconductor technology. This economic aspect is of course important for the production of low-cost, disposable sensors. The disadvantage of the thick-film technology is the larger dimension of the device because of the minimal line-width of 100 μm. This implies, for example, that thick-film sensors are not suited for applications where extreme miniaturization is necessary (e.g. 'in vivo' sensors).

Thick-film sensors are mainly used in the automotive industry for the measurement of pressure and force and for combustion control (Heintz and Zabler 1987, Prudenziati 1987, 1991). In the chemical field, a great variety of papers can be found on thick-film gas sensors based on SnO_2 and lean-combustion sensors based on ZrO_2. Until recently, very few papers could be found on the use of thick-film technology for the preparation of classical electrochemical devices such as potentiometric or voltammetric sensors, for example. In Pace and Jensen (1986) and in Belford *et al* (1987), a thick-film pH sensor is described. In Pace and Jensen (1986), a screen-printed PVC/ionophore layer is used as a pH-sensitive element. Belford *et al* (1987) uses classical pH-sensitive glass mixtures, screen printed on a multi-layer metal conductor. In Pace *et al* (1985), a thick-film multi-layered oxygen sensor with screen-printed chemical membranes in PVA and silicone rubber is detailed (a similar approach is described by Karagounis *et al* (1986)). No chemically active layers, however, were applied. An electrocatalytic glucose sensor is presented in Lewandowski *et al* (1987). At ESAT–MICAS, RuO_2 electrodes have been developed for use in a voltammetric biosensor (Suls *et al* 1986). An enzyme-based glucose sensor with RuO_2 electrodes is detailed in Lambrechts *et al* (1987). All these thick-film sensors are fabricated on Al_2O_3 substrates. In Weetall and Hotaling (1987), an extremely low-cost, immuno-assay sensor is presented which is silk-screen printed on cardboard. Some of these developments are also discussed in chapter 6. In Cha *et al* (1990b), the perfomance of thick-film Au and Pt electrodes is compared with conventional electrodes.

Since thick-film technology is a microelectronic technique for the fabrication of small, high-performance electronic circuits, smart thick-film sensors can be made easily by incorporating some interface electronics onto the same substrate using surface-mount technology (SMT) (Seipler 1987). Thick-film technology is especially interesting for the calibration of silicon sensors by trimming resistors on the sensor substrate. This technique is used commercially for pressure sensors and accelerometers (NovaSensor 1987, IC Sensors 1987). For chemical sensors, no commercial applications of smart thick-film sensors are known.

In Atkinson *et al* (1991), three seperate thick-film circuits are described: a potentiostat, a ramp genrator and the thick-film sensor itself. As they are not combined on a single susbstate, we can not state that this system is a smart thick-film sensor. The possibility of making smart thick-film sensor is, however, self-evident.

It can be concluded that thick-film technology is an exciting alternative to CMOS- or bipolar-compatible sensors. They are suitable for applications where extreme miniaturization is not necessary. In combination with surface-mounted devices, smart sensors can be made with commercially available components. Screen printing is also an interesting technique for the application of planar chemical membranes onto silicon substrates. Therefore, a major breakthrough in the field of thick-film sensors is expected within the next five years.

4.6 CONCLUSION: A COMPARISON BETWEEN DIFFERENT SENSOR TECHNOLOGIES

The choice between different sensor technologies is mainly an economic decision. According to the application and the market for a certain sensor product, the manufacturing technology has to be determined. The main question is: 'How many sensors can I produce and market for what price?'.

An answer to that question is presented graphically in figure 4.43. A low sensor price can only be achieved with high production quantities. For high-volume production, a reliable mass-production technique is needed. The choice between thick-film, thin-film or IC technology depends on the total number of sensors per year that can be marketed. It is absurd to use IC technology for a sensor market where only a thousand sensors per year

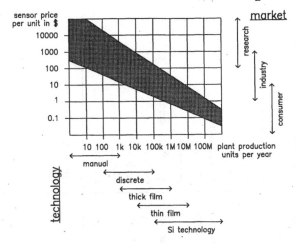

Figure 4.43 The economic aspects of sensor technology: price, market and technology (adapted from Heintz and Zabler (1987)).

are needed. Mass-production techniques are only viable when the total number of plant production units per year are sufficiently high; this is the only way to write off the high initial investments.

Another important aspect of sensor price is the sensor lifetime. A sensor can be developed with two different objectives: high-priced sensors with a long lifetime or low-priced, disposable sensors. Disposable sensors are the ideal sensors for mass-production techniques.

In table 4.3 the different aspects of sensor technology are summarized into three categories: production techniques, economic aspects and typical characteristics.

The manual production of sensors is an interesting approach in the early development of a new sensing technique, as the prototyping cost and the initial investment are low. The market for these high-priced sensors is limited to research and aerospace applications.

Thick-film technology is a low-cost mass-production technique. Due to the larger dimensions, the market is restricted to automotive and industrial applications. The reproducibility of thick-film sensors can be increased by using trimming techniques. Advantages of thick-film technology are the flexibility of the process and the low cost of prototyping. The polymer-based systems offer great advantages (not mentioned in table 4.3). Polymer thick-films can be screen printed on cheap polymer substrates and can be produced in significant quantities (> 1 million units per year) at a low cost. They are mainly suited to disposable sensor applications.

Thin-film technology on glass or ceramic substrates uses the same sensor fabrication as for ICs. It has some of the advantages of thick-film technology, but lacks the advantages of IC technology. The process flexibility is of the

Table 4.3 A comparison between different sensor technologies: economic and technical aspects.

	classic construction	thick film technology	thin film technology	IC technology
technology	wires and tubes	screen printing	evaporation-sputtering	IC techniques
substrate		Al_2O_3, plastic	Al_2O_3, glass, quartz	silicon, GaAs
initial investment	very low	moderate	high	high
production line cost	> 10 k$	> 100 k$	> 400 k$	> 800 k$
production	manual production	mass production	mass production	mass production
units per year	1 - 1000	1.000 - 1.000.000	10.000 - 10.000.000	100.000 - ...
prototype	cheap	cheap	moderate	expensive
sensor price	expensive sensor	low cost per sensor	low cost per sensor	low cost per sensor
use	multiple use	disposable	disposable	disposable
	in vitro - in vivo	in vitro	in vivo	in vivo
markets	research	automotive	industrial	medical
	aerospace	industrial	medical	consumer
dimension	large	moderate	small	extreme miniaturization
solidity	fragile	robust	robust	robust
reproducibility	low	moderate	high	high
max. temperature	800 °C	800 °C	1000 °C	150 °C (Si)
interfacing	external	smart sensors	smart sensors	smart sensors
	discrete devices	surface mount	surface mount	CMOS, bipolar

same level as for ICs. Taking into account that nowadays silicon foundries deliver processed wafers with custom-designed interface electronics at a reasonable cost and in a reasonable time, the use of thin-film technology is only justified when no on-chip interface electronics are necessary.

IC fabrication, and especially CMOS or bipolar technology, is without doubt an interesting basis for the production of cheap, disposable sensors. Interface electronics can be easily integrated with the sensor. The resulting devices are extremely small so that they can be used for 'in vivo' medical applications. For some applications the maximum allowed temperature of 150 °C can be a problem.

The availability of specific microelectronic-compatible sensor technology is the prime difficulty of IC technology based sensors. The low price, comparable with the price of an ordinary integrated circuit, can only be achieved when every processing step, including packaging, is based on microelectronic mass-production techniques. The next challenge is the marketing of these huge quantities of planar sensors.

These economic factors have been taken into account in this book. The development of both thick-film and CMOS-compatible sensors is described: with these two mass-production techniques a broad range of sensors can be produced.

Attention is also paid to the conversion of microelectronic techniques to sensor-specific processing, as described in this chapter. Different lift-off techniques are described and are used in a generic CMOS-compatible process (see chapter 5). For the patterning of chemically active membranes, photolithography, lift-off, plasma etching and screen printing are proposed.

CHAPTER 5

CASE STUDIES ON MICROELECTROCHEMICAL SENSORS

The theory and technology required to fabricate planar electrochemical sensors have been detailed in the previous sections. In this chapter, practical realizations of sensors developed by the authors, as well as by other research groups, are discussed. Although this book covers mainly biosensors, some electrochemical bioprobes will also be presented. Indeed, an identical technological basis provides the facilities for the realization of biosensors as well as bioprobes.

5.1 A GENERIC, CMOS-COMPATIBLE PROCESS FOR ELECTROCHEMICAL SENSORS

In this section, the development of a novel, generic CMOS-compatible process for electrochemical electrode systems is described. Also, several new planar voltammetric sensor designs are discussed. It is important for commercial sensor production to have a process with high flexibility so that the sensor characteristics can be adapted easily to the specific needs of a new application. With this option in mind, a new process for planar voltammetric glucose sensors has been developed.

It is a generic process method; this means that the techniques used can also be applied to other electrochemical sensor types, such as planar oxygen sensors, ion-selective potentiometric sensors, gas sensors or conductometric cells. This generic nature of the process is based on CMOS compatibility and the CVD oxide assisted lift-off technique described in section 4.2.3. This universal characteristic is unquestionable for the electrode system itself; with a voltammetric electrode system for H_2O_2, a broad range of biosensors can be made by changing the active enzyme in the enzymatic membrane. However, universal coating techniques for chemically active membranes do not

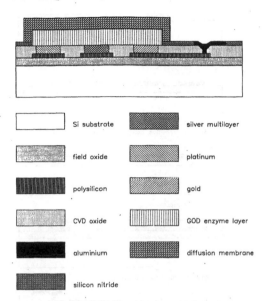

Si substrate	silver multilayer
field oxide	platinum
polysilicon	gold
CVD oxide	GOD enzyme layer
aluminium	diffusion membrane
silicon nitride	

Figure 5.1   The cross section of a CMOS-compatible sensor.

exist (see section 4.4) because of their specific characteristics. This specificity interferes with the development of generic methods for the patterning of chemically active membranes.

In figure 5.1, the cross section of the CMOS-compatible sensor structure is shown. As many layers as possible from the CMOS process are used. The CMOS layers, such as field oxide, polysilicon, CVD oxide and aluminium, are already necessary for the on-chip interface electronics. By using these standard layers, sensor processing is reduced, resulting in a lower cost.

The field oxide and the CVD oxide are used as isolation layers between the substrate and the electrodes. Interconnection of the noble-metal electrodes and the interface electronics or the aluminium bonding pad is achieved with the standard polysilicon layer. Polysilicon is preferred to diffused layers for its lower resistivity and light insensitivity. As a passivation layer, silicon nitride is used. This passivation is necessary to protect the aluminium metallization pattern from corrosion.

5.1.1   The different process steps

The sensor processing can be divided into two distinct parts; the CMOS part and the sensor part. CMOS technology is very widespread, so this part of the process can be carried out in any silicon foundry throughout the world. The basic steps of a CMOS process are described in section 4.1.1. If no electronics are needed on the sensor, the CMOS process can be simplified to the following steps (figure 5.2); these steps are identical to those used

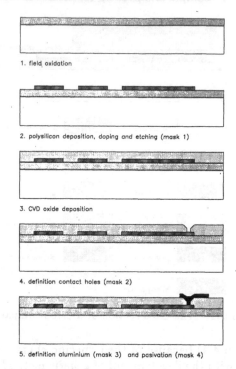

1. field oxidation

2. polysilicon deposition, doping and etching (mask 1)

3. CVD oxide deposition

4. definition contact holes (mask 2)

5. definition aluminium (mask 3) and pasivation (mask 4)

Figure 5.2   The different steps of the simplified CMOS process.

in the LUVCMOS 3 process developed at K U Leuven ESAT (K U Leuven ESAT 1983).

1. Cleaning of the wafers

2. Field oxidation

The field oxide is a wet thermal oxide; the wafers are oxidized in a steam atmosphere for 10 hours at 975 °C. This results in a 1.2 μm thick field oxide.

3. polysilicon deposition, doping and patterning (mask 1)

The polysilicon layer is deposited in an LPCVD furnace at a temperature of 630 °C. The thickness is 0.5 μm. After deposition, the polysilicon layer is doped with phosphorus using a solid-state source. The sheet resistivity of the layer is typically 20 Ω/square. After doping, the polysilicon layer is wet etched with 'POLY' etch (1250 ml HNO_3 65%, 500 ml H_2O, 37 ml BHF).

4. CVD oxide deposition

The CVD oxide is deposited in two steps. Firstly, two layers are deposited. The first layer (0.2 μm) is a pure CVD oxide and the second layer (0.8 μm) is doped with phosphorus (10%), so a phosphosilicate glass (PSG) with a low melting point is formed. After a first etch of two thirds of the CVD oxide in the contact holes, the CVD oxide is reflowed in order to decrease the slope. Thereafter, an undoped layer (0.2 μm) is deposited.

5. Etching of the contact holes (mask 2)

The contact holes are etched in two steps, as indicated above. The first time, two thirds of the thickness of the first two CVD oxide layers are etched; the second time, the upper CVD oxide layer and the remaining third are removed in the contact regions.

6. Deposition of Al/1%Si, patterning and sintering (mask 3)

After cleaning the wafers and after a HF 5% dip to remove the native oxide in the contact holes, a 1 μm thick aluminium layer is sputtered using a DC magnetron plasma-sputtering machine. To decrease the diffusion of the aluminium in the silicon, an Al/1%Si alloy is used. The aluminium layer is then patterned by wet etching. After cleaning, the wafers are sintered in forming gas for 30 minutes at 450 °C to decrease the contact resistance.

7. Plasma nitride deposition and patterning (mask 4)

A 0.5 μm plasma silicon nitride layer, followed by a thin (0.2 μm) CVD oxide, is deposited. The CVD oxide is necessary to improve the adhesion of subsequent polymer membranes to the substrate; this layer is not necessary for passivation of the electronic structure and is not standard for a normal CMOS process. This multilayer is then etched away from the sensor area and the bonding pads using dry-etching techniques. Most sensors described in this book were not passivated, as there were no interface electronics on the chip. Passivation is only necessary for long-term experiments in this case.

These seven steps of CMOS processing, using four masks, are necessary to prepare the CMOS layers used in the sensor structure; a complete CMOS process is built up in forty steps using typically twelve masks (K U Leuven ESAT 1983). This complete process is, of course, only required when on-chip interface electronics are required.

Whereas the first (CMOS) part is based on known, standard CMOS technologies, the second (sensor) part is based on new sensor techniques developed by the authors and described in the previous chapter. In the second part of the process, the specific sensor layers are applied on the finished CMOS wafer. Since the sensor processing follows on from the finished CMOS process, no high-temperature steps are allowed. The maximum temperature during sensor processing is 400 °C; above this temperature the aluminium will diffuse into the source and drain junctions, causing a short circuit. This limits the applicable sensor technology to low-temperature methods, such as photolithography, evaporation, sputtering, LTCVD and wet and dry etching. All high-temperature steps (oxidation, diffusion, annealing, LPCVD) are excluded. Since annealing of the structures is no longer possible, radiation damage during sensor processing has to be avoided.

The sensor process varies according to the specification of the sensor (e.g. glucose sensor or oxygen sensor). The main building blocks of an electrochemical sensor, however, remain the same: noble-metal electrodes, an electrolyte layer and a chemically sensitive membrane. In this generic process, the noble-metal electrodes are applied with a universal lift-off method so that the basic processing remains the same for the complete range of electrochemical sensors. As an example, the processing of a three-electrode oxygen/glucose sensor is detailed. For an oxygen sensor, an Ag/AgCl reference electrode, a Pt auxiliary electrode and an Au working electrode are needed. This electrode structure is then covered with a hydrogel layer (the electrolyte) and an oxygen diffusion membrane. In the process developed by the authors, five masks are needed for this purpose. An overview of the different sensor process steps for an oxygen sensor are given in figure 5.3. For a glucose sensor, an Ag/AgCl reference electrode and Pt auxiliary and working electrodes are needed. This electrode structure is then covered with an enzymatic membrane and a diffusion membrane. For oxygen and glucose sensors, the same electrode configuration and mask set are used. In comparison with an oxygen sensor, a glucose sensor has a Pt working electrode and the membranes are also different. Therefore, only four sensor-specific masks are necessary for a glucose sensor.

1. Lift-off of the silver layer (mask 5)

The silver layer and all other noble-metal layers are patterned with the CVD oxide assisted lift-off method described in section 4.2.3. The silver is deposited on the wafer as a Ti/Pd/Ag layer using electron-gun evaporation. This multilayer is used for its excellent corrosion characteristics (see section 5.2.4). The Ag layer can be converted into Ag/AgCl in a 1% $FeCl_3$ solution before the lift-off of the metal. As an alternative, the electrodes can be chloridated afterwards on the individually mounted sensors or on the wafer using photoresist as a masking material.

2. Lift-off of the platinum layer (mask 6)

The platinum electrodes are also patterned with the same CVD oxide assisted lift-off method. For good adhesion, a thin titanium layer is sputtered before the sputtering of the platinum layer. During the sputtering process, special attention has to be paid to avoid increasing the wafer temperature.

3. Lift-off of the gold layer (mask 7)

The gold electrodes are also patterned with the CVD oxide assisted lift-off method. For good adhesion and corrosion resistance, a Ti/Pd/Au layer is used. Firstly, the Ti/Pd layers are evaporated onto the substrate using electron-gun evaporation, followed by the Au layer using resistive evaporation. This Au layer is not needed for glucose sensors.

4. Spinning and patterning of the hydrogel (mask 8)

The photosensitive hydrogel material (PVA, PVP or PHEMA) is spin coated onto the wafer. With UV exposure, the material is cross-linked. During the development in lukewarm water, the unexposed polymer is removed.

For glucose sensors, this membrane is replaced with a glucose oxidase membrane. GOD membranes based on an albumin matrix can be patterned

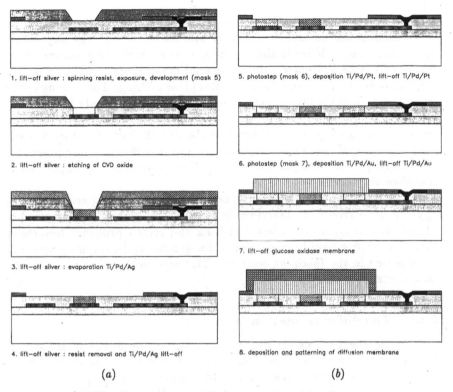

1. lift-off silver : spinning resist, exposure, development (mask 5)

5. photostep (mask 6), deposition Ti/Pd/Pt, lift-off Ti/Pd/Pt

2. lift-off silver : etching of CVD oxide

6. photostep (mask 7), deposition Ti/Pd/Au, lift-off Ti/Pd/Au

3. lift-off silver : evaporation Ti/Pd/Ag

7. lift-off glucose oxidase membrane

4. lift-off silver : resist removal and Ti/Pd/Ag lift-off

8. deposition and patterning of diffusion membrane

(a)   (b)

Figure 5.3   The different steps of the sensor process.

with the lift-off technique described in section 4.4.4. The same mask can be used for this purpose.

5. Spinning and patterning of the diffusion membrane (mask 9)

As a diffusion membrane for an oxygen sensor, a normally available photoresist can be used (Suzuki *et al* 1988). A possible alternative is dry etching of, for example, teflon membranes. A new amorphous fluoropolymer (TEFLON AF) has been developed by Dupont and is possibly of interest for this application. This new teflon material can be dissolved in special solvents and can be applied by spin coating (Dupont 1989).

For the diffusion membrane of glucose sensors (cellulose acetate or polyurethane), no good patterning techniques are known; screen printing and dry etching seem to be valid alternatives for the CMOS-compatible processing of these membranes.

It can be concluded that with this CMOS-compatible process sequence, a great variety of electrochemical sensors can be realized. Because of the universality of the CVD oxide assisted lift-off process, a generic sensor technology has been developed with respect to the noble-metal electrodes. The etching of contact holes, followed by the etching of the metal layer, is an alternative solution to this lift-off process. However, with this method an extra mask is needed for the etching of the contact holes. The step coverage in the contact holes is also a problem, as no high temperature reflow of the PSG layer is allowed. Hence, special dry-etching techniques with a sloped profile are necessary for the contact hole etching. The etching of the noble-metal layers is also difficult. For example, wet etchants for Pt exist, namely aqua regia, but are not compatible with standard resists. Plasma-assisted etching of noble-metal layers is also not obvious. The best method for the etching of noble-metal layers is ion-beam milling; the equipment for this technique, however, is not yet in a mature commercial phase and is therefore very expensive.

For specific chemical membranes, no universal CMOS-compatible technology can be derived. As described in section 4.4, possible techniques are photolithography, lift-off, plasma etching and screen printing. Several of these techniques have been developed for glucose and oxygen sensors and are detailed in the following sections. This section deals mainly with the bare electrode system.

5.1.2   The different mask sets

The development of planar sensors also involves the design of the sensor layout, i.e. the mask art-work. In figure 5.4, the basic layout of a three-electrode voltammetric sensor is shown. Whereas the major concern in the development of sensor processing involves the cross section of the device, sensor layout is basically considered from a top view. The processing sequence and the sensor layout are nevertheless strongly linked, as each processing layer requires a different mask. The critical dimensions on the mask (layout rules) are also defined by processing limitations.

During the layout, the electrode geometry is defined. No relevant publications are known on the influence of planar electrode geometry on sensor response and performance, so most sensor layout is done on an intuitive basis.

To eliminate this trial-and-error method, a theoretical study of the influence of electrode geometry on sensor performance has to be undertaken. However, as this problem is mathematically very complex (see section 2.3.1), numerical solutions have to be developed. A possible alternative is the evaluation of electrode geometry with finite-element analysis. Such theoretical studies, in combination with experimental verification, could then lead to the necessary CAD (computer aided design) tools for automated sensor design. Nowadays, no CAD tools for chemical sensor design are available or reported in the literature. As there is an extensive need for these, research on this topic is recommended.

5.1.2.1 Layout rules for sensor design. The following constraints and

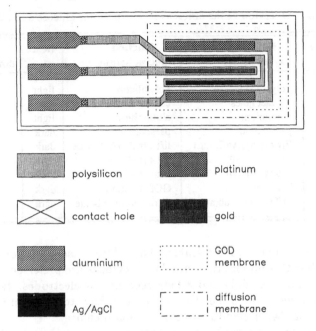

polysilicon

platinum

contact hole

gold

aluminium

GOD membrane

Ag/AgCl

diffusion membrane

Figure 5.4   The basic layout of a three-electrode voltammetric sensor.

guide-lines have to be taken into account for the sensor layout.

1. Sensor dimensions

The dimension of the sensor is limited to a well defined area. For economic reasons, the area of the sensor should be minimized, since, in the microelectronic world, price is proportional to the total silicon area.

For 'in vivo' applications, where mounting in a catheter is necessary, the width of the sensor is important. For mounting in a 1.65 mm external diameter catheter (French 5), the width of the sensor must not exceed 0.7 mm. When extreme miniaturization of the catheter is necessary, these dimensions must even be reduced to 0.4 mm for a 0.94 mm catheter (French 2.8).

2. Electrode configuration

During the layout phase, the electrode configuration is defined. No information can be found on this topic in the literature. As a general rule, symmetrical configurations are preferred.

3. Electrode area

The current of a voltammetric sensor is proportional to the area of the working electrode. However, response time and flow dependence increase with increasing electrode area, so small working electrodes are preferred. Noise considerations, however, limit the possible miniaturization (see section 5.2.4). The $1/f$ noise is inversely proportional to the electrode area,

Table 5.1   Overview of the different masks used for glucose and oxygen sensors.

number	function oxygen sensor	function glucose sensor	field	critical dimension μm
1	polysilicon	polysilicon	light	5
2	contact holes	contact holes	dark	5
3	aluminium	aluminium	light	5
4	passivation	passivation	dark	10
5	lift-off Ag/AgCl (RE)	lift-off Ag/AgCl (RE)	dark	5
6	lift-off Pt (AE)	lift-off Pt (AE)	dark	5
7	lift-off Au (WE)	lift-off Pt (WE)	dark	5
8	hydrogel	GOD membrane	dark	10
9	diffusion membrane	diffusion membrane	dark	10

so small electrodes are less stable. The white noise is proportional to the form factor; this implies that circular electrodes are preferred above square electrodes and certainly above long rectangular electrodes. However, in a practical design, where the other guide-lines have to be taken into account, long rectangular electrodes are the optimum choice.

The reference electrode of the sensor can be designed with the same limitations as described above. The auxiliary electrode has to be larger than the working electrode to prevent current limitation at this electrode. According to the literature (Bard and Faulkner 1980), the area of the auxiliary electrode should be, in general, three times larger than the area of the working electrode. A factor of 1.5 is, however, sufficient for the applications anticipated in this book.

4. Electrode spacing

Very little is known on electrode spacing. A close spacing of the different electrodes should be avoided to prevent migration and cross-influence of the electrodes. In the designs presented in this book, a minimum electrode spacing of 50 μm is used; a larger spacing was used when possible. A recent study in (Cha *et al* 1990a) indicates that chemical cross-talk among electrodes is possible if the electrode spacing is reduced.

5. Distance between electrodes and bonding pads

To allow easy packaging of the sensor, the distance between the chemically active electrodes and the bonding pads must be sufficiently large. In the designs presented by the authors in this book, a minimum distance of 1 mm is used, where possible.

6. CMOS layout rules

Since this sensor processing is based on CMOS technology, it is self-evident that the layout rules for CMOS design have to be followed. These design rules are always made available to the circuit designer by the silicon foundry.

The different masks and the process sequence are sketched in figure 5.5.

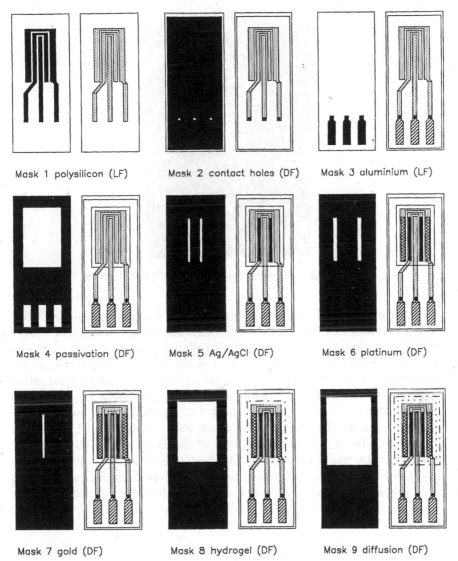

Figure 5.5   The different masks and the process sequence of the basic electrochemical sensor (figure 5.4).

For a CMOS-compatible oxygen sensor without interface electronics, nine different mask are necessary: four CMOS processing masks and five sensor processing masks. An overview of the complete mask set is given in table 5.1. With this schematic mask set, both oxygen and glucose sensors can be made.

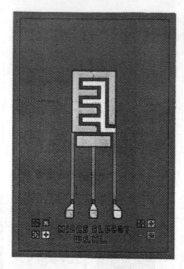

Figure 5.6   Microphotograph of the 'GLUCO' sensor.

5.1.2.2 'GLUCO'. The mask set 'GLUCO' was designed at the beginning of 1984, during the first year of electrochemical sensor activity at ESAT-MICAS. At that time, the knowledge of planar chemical sensors was limited. So, the design of this sensor was inspired by the knowledge of glucose sensors at the Université Libre de Bruxelles (Garcia *et al* 1983). This rather large sensor (2.6 mm on 3.9 mm) was the first-generation prototype of a planar CMOS-compatible glucose sensor (see figure 5.6). Extreme miniaturization was at that time not the main issue, as, first of all, the sensor process had to be optimized. Because of the large dimensions, only one sensor per die was implemented. No test structures for in-process testing were provided. At that time, a finger-structure electrode configuration was proposed for the working and reference electrode, combined with a small auxiliary electrode. After evaluation it was seen that this electrode configuration has several important drawbacks.

1. The auxiliary electrode is too small in comparison with the area of the working electrode. Theoretically, the dimension of the auxiliary electrode does not influence the response of the sensor. However, a small electrode implies a high current density at this electrode. For planar structures with thin electrodes, high current densities should be avoided as these currents destroy the electrode. As a result of this small auxiliary electrode, the lifetime of the 'GLUCO' sensor is limited due to loss of adhesion of the auxiliary electrode.

2. The finger-structure electrode configuration has been chosen in order to obtain a uniform potential distribution around the working electrode. However, since physiological solution has a high conductivity, this is not

necessary. Because of the high solution conductivity, the potential in the solution is uniform at a distance of a few microns from the reference electrode.

Also, this finger-structure electrode configuration is not appropriate from the migration viewpoint. Migration of Ag/AgCl from the reference electrode to the working electrode is enhanced by a non-uniform current distribution. In this finger structure, several points with a high current density exist. Therefore, an electrode configuration with a symmetrical and uniform current distribution is preferred.

3. The sensor is too big to fit into a catheter.

4. The bonding pads are too small to allow the attachment of wires to the sensor with alternative techniques; only wire bonding can be used.

5. There are no test structures provided.

6. In this layout, the passivation layer overlaps the noble-metal electrodes. This implies that passivation is the last step in the process sequence. In that case, Si_3N_4 cannot be used due to temperature restrictions and the risk of contaminating deposition equipment. The adhesion of other passivation layers, such as polyimide, to noble metals is bad. Therefore, the active electrode area is not well defined.

These drawbacks have been corrected in the following layouts. Nevertheless, the 'GLUCO' mask set was essential for testing the sensor process and has been used with success for processing voltammetric sensors with a great variety of electrode materials. Although this layout is far from excellent, an identical layout is used by other researchers in a European patent application (Otagawa and Madou 1989) without reference to the papers published earlier by the authors.

5.1.2.3 'VOLTA I'.   The 'VOLTA' series is a series of designs especially suited for catheter-tip packaging. The 'VOLTA I' mask set was designed in 1985 when it became clear that the 'GLUCO' set had too many drawbacks. The major improvements of the 'VOLTA I' set are the larger auxiliary electrode, the linear electrode configuration and the smaller dimensions.

In the 'VOLTA I' mask set the area of the auxiliary electrode is made 1.4 times larger than the area of the working electrode in order to prevent high current densities at the auxiliary electrode. The working electrode has been made smaller to decrease the flow dependence of the sensor.

Four different linear electrode configurations are placed on one die (figure 5.7). The linear electrode configuration has been chosen for the symmetrical potential and current distribution.

The first two configurations differ in the sequence of the electrodes. For the type 1 sensor the sequence is AE-RE-WE; for the type 2 sensor the sequence is AE-WE-RE. Sensor types 3 and 4 are the symmetrical versions of 1 and 2, respectively. For the type 3 sensor the sequence is AE-RE-WE-RE-AE; for the type 4 sensor the sequence is AE-WE-RE-WE-RE.

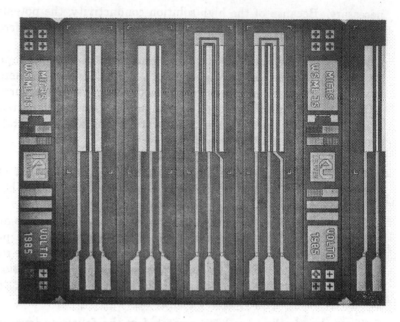

Figure 5.7   Microphotograph of the 'VOLTA I' sensor.

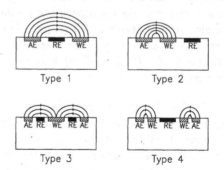

Figure 5.8   Ion current distribution at different 'VOLTA I' sensors.

The sensors also differ in the current path between the different electrodes. In figure 5.8 the supposed current path is sketched for the different electrode configurations. Out of these presumptions it can be concluded that the type 3 and 4 electrode configurations are the most interesting as they are symmetrical.

In practice, no difference has been found in the electrochemical response of type 1, type 2 and type 3. The type 4 electrode configuration gives a larger faradaic current, as the working electrode is split into two small electrodes. As discussed in section 3.2.6, the sensitivity of a band electrode increases with decreasing electrode width. It was expected that a differ-

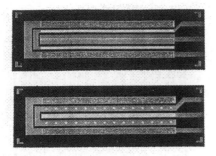

Figure 5.9   Microphotograph of the 'VOLTA I' sensor structure with ultra-microelectrodes.

ence in lifetime between the different configurations would manifest itself because of migration problems. However, no experimental confirmation of this presumption has been found.

The bonding pads of the 'VOLTA' sensors have been made longer in compared to the 'GLUCO' layout. This allows the fixing of wires to the sensor with techniques other than wire bonding. The dimensions are suited to the fixing of isolated copper wires by thermocompression or point welding. The dimensions of the bonding pads and of the sensor itself are compatible with the catheter-tip packaging line at Dräger Medical Electronics (Best, Netherlands). This catheter-tip packaging is necessary for 'in vivo' use of the 'VOLTA' sensor.

On the 'VOLTA' layout, test structures are provided for the control of critical dimensions during all the different photolithographical steps. There is also a contact test structure to check the standard CMOS processing.

The 'VOLTA I' mask set has been used with success for the processing of voltammetric sensors with a great variety of electrode materials. A variant of the structures shown in figure 5.7 has been implemented for a study of the behaviour of ultra-microelectrodes. Therefore, a 10 μm wide band electrode and a multi-microelectrode (with 28 circular electrodes with a diameter of 10 μm) have been realized, as shown in figure 5.9.

5.1.2.4 'VOLTA II'.   The 'VOLTA II' mask set is the mature design of a voltammetric glucose or oxygen sensor (see figure 5.10). On one die, a two-electrode and a three-electrode oxygen sensor are combined with a three-electrode glucose sensor. This sensor design is compatible and identical with the sensors on the CMOS interface chip described in section 5.3.4. The purpose of this design was the testing of the electrochemical characteristics of the sensor on the interface chip without the use of on-chip interface electronics. In this way the sensor design could be evaluated before the integration of the sensor on the CMOS interface chip itself.

The layout of the sensor is almost identical to the 'VOLTA I' design. The working electrodes are made smaller in order to diminish the flow

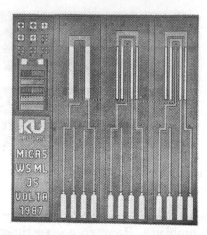

Figure 5.10   Microphotograph of the 'VOLTA II' sensor

dependence. The width of the working electrode of the oxygen sensors is 10 μm; the width of the working electrode of the glucose sensor is 20 μm. With these dimensions, the minimum achievable electrode width is reached. Smaller electrodes can be prepared with lift-off techniques, but the reproducibility of these electrodes would be rather low. The current response at the 'VOLTA II' electrodes is also very low. For an oxygen sensor a current of 20 nA is measured for 25 kPa pO_2. It becomes difficult to measure these currents with an acceptable precision without using costly equipment. The use of on-chip interface electronics becomes necessary for these scaled sensors. The corresponding CMOS on-chip interface electronics developed in ESAT–MICAS are discussed in section 5.3.4.

As can be seen in the layout of figure 5.10, a type 3 electrode configuration has been implemented. A small improvement has been made on this configuration; the working electrode is longer than the auxiliary and reference electrode, improving the migration characteristics of the sensor. Without this elongation, a higher current density exists at the extremities of the working electrode.

5.1.2.5 Concluding remarks on voltammetric sensor layout.   During the years of our research, the sensor layout has evolved from a crude, imperfect prototype to a mature design. It can be concluded, however, that the design of these planar electrochemical sensors resulted more or less from a trial and error method. In the world of microelectronic circuit design, this is an unusual practice, as all circuits nowadays are simulated and tested with CAD tools before processing. However, these are not available for electrochemical sensor design. As there is an extensive need for these CAD tools, research on this topic is recommended.

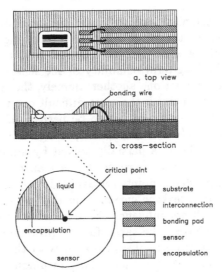

Figure 5.11   Schematic cross section of the packaging of an electrochemical sensor.

5.1.3   Packaging

The protection of a sensor is sited on two different levels. First of all, the passivation layer protects the electronic circuits and interconnections on the sensor (see section 4.1). As passivation is a wafer-level process, the cost of good passivation is relatively low. Packaging, on the other hand, protects the interconnections between the sensor and the external interface circuits. It is evident that packaging and passivation have to be tuned very accurately. Packaging is done on the individual sensors and is thus an expensive process step; the cost of packaging planar sensors is high in comparison with the fabrication cost of the sensor die itself. This cost aspect is enhanced even further by the fact that it is very difficult to automate the packaging of sensors; currently this is still an expensive manual process.

Packaging of a planar sensor is one of the most difficult tasks in the development of reliable electrochemical sensors. The goal is to exclude interference from the hostile environment. These interferences can be, for example, corrosion of the sensor materials and leakage currents between high-impedance points on the sensor or the interface electronics. Packaging of chemical sensors is therefore a contradiction in terms, since the chemical part of the sensor has to be exposed to the environment and the electronic part has to be completely protected from the same environment. From this statement it can be concluded that hard-shell encapsulation (such as titanium packages for cardiac pace-makers) cannot be used for the packaging of planar chemical sensors.

Packaging of a chemical sensor also implies, as indicated in figure 5.11, that there is a zone on the sensor where the packaging stops and the chemically active site of the sensor begins. The stability of this zone is of critical importance for the long-term reliability of the sensor. In this zone, three different materials contact one another; namely, the sensor, the liquid and the encapsulant. This point is the most difficult one to control as adhesion of the encapsulant to the sensor has to be perfect. If an encapsulant with a low moisture permeability is used, the package will start to deteriorate.

The packaging problem is partially solved by adding on-chip interface electronics to the sensor; a smart sensor is realized and internal leakage currents at high-impedance points are avoided by passivation of the circuit with Si_3N_4. External leakage currents on the low-impedance outputs of the smart sensor will not influence the response of the sensor. However, special precautions have to be taken for the protection of the sidewalls of the chip, as the substrate is always biased with the positive or negative power supply voltage. A novel packaging technique for smart sensors is proposed in section 5.1.3.

Although the packaging problem is generally cited in the literature, no simple, efficient packaging methods are described. Most of the packaging methods reported are based on epoxies or silicone rubber for the protection of the device (Sibbald 1985, Koudelka 1985). Only recently have packaging methods based on micromachined glass or silicon been detailed (Blenneman 1987, van den Vlekkert 1987, Smith 1988). As will be discussed further, these methods promise to have interesting characteristics for the long-term packaging of electrochemical sensors.

5.1.3.1 Epoxies. Epoxies are frequently used for the packaging of chemical sensors. They consist typically of two components: the resin and the hardener. A quick temperature cure is preferred for enhanced performance of the epoxy. Well known epoxies for the packaging of sensors are described in the literature, namely Epo-tek H54, Araldite and UHU plus. A real biocompatible epoxy does not exist at this moment. In practice, the best results obtained by the authors have been with the expensive Epo-tek H54 epoxy. The problem with this material, however, is the high-temperature cure. This epoxy cannot be used for the packaging of temperature-sensitive sensors such as enzyme-based biosensors. In Dumschat *et al* (1990), a method for the direct photolithographic structuring of epoxy resins is demonstrated for the encapsulation of CHEMFETs.

A general problem with epoxies is associated with the required mixing procedure of a two-part system. Even with very careful mixing of the two components, entrapment of air bubbles in the mixture cannot be avoided. Therefore de-aeration in vacuum is required. This is a tiresome procedure. It is also of great importance to arrive at a stoichiometrically balanced mixture of the two components. The presence of excess resin or hardener seriously undermines the physical properties of the encapsulant.

5.1.3.2 Silicone rubbers. Silicone rubbers are known for their good moisture resistance and for their excellent adhesion on glass-like material. This superior adhesion of silicone rubber over epoxies and other polymer encapsulants is due to the fact that silicone rubber remains both an elastomer and an adhesive in the presence of water. The water permeability and absorption of silicone rubbers is nevertheless higher than for epoxies. It was shown by Donaldson (1976) that the high water absorption and permeability of a polymer does not degrade its ability to prevent corrosion and leakage currents, as long as it maintains adhesion to the substrate. More recent studies correlate this behaviour with osmosis effects (Donaldson 1991). Maintaining adhesion prevents water vapour from condensing in voids at the substrate–polymer interface, so the formation of liquid is prevented; this liquid is required for ionic leakage currents and corrosion. Therefore, silicone rubbers are sometimes preferred above epoxies.

Silicone rubbers are typically single-component materials. Curing is accomplished at room temperature and is initiated by the moisture in the air. These silicone rubbers are designated as RTV (Room Temperature Vulcanize) materials. After curing, a soft clean product is formed. Some silicone rubbers are also biocompatible. These are usually based on two components and the hardening can be accelerated by a temperature cure.

Several types of silicone rubbers have been tested in this study. It was concluded that Dow Corning Conformal Coating R-4-3117 is a good encapsulant. It has a low water permeability and is suitable as a secondary encapsulant for chemical sensors. In Devanathan and Carr (1980) it is concluded that this coating is second best, after Parylene C, for the encapsulation of pace-makers.

5.1.3.3 Parylene C. Parylene is the generic name for a series of thermoplastic polymers developed by Union Carbide. The basic polymer is Parylene N or poly-para-xylylene. The best parylene for encapsulation is Parylene C or monochloro-para-xylylene. The structure of both polymers is shown in figure 5.12. Parylene C exhibits superior dielectric strength, exceptionally high surface and volume resistivities and low permeability to moisture and gases. According to the claim of Union Carbide, Parylene C is biocompatible and non-thrombogenic. It is insoluble in conventional solvents and has to be applied on the devices in a special vacuum pyrolysis system.

Because of the low pressure during the deposition, a uniform conformal coating is obtained. The typical thickness of this coating is between 1 μm and 5 μm. Adhesion of Parylene C to a rough surface is excellent. However, as no chemical bond is formed between the material to be covered and the

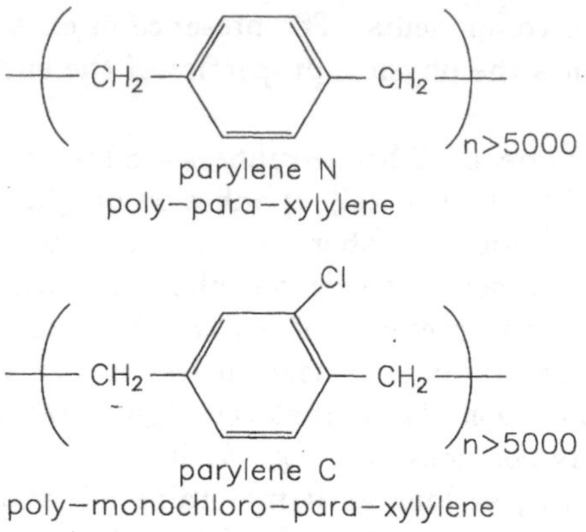

$$\left(- CH_2 -\bigcirc- CH_2 -\right)_{n>5000}$$

parylene N
poly−para−xylylene

$$\left(- CH_2 -\bigcirc- CH_2 -\right)_{n>5000}$$

parylene C
poly−monochloro−para−xylylene

Figure 5.12   Chemical structure of Parylene N and Parylene C.

parylene layer, adhesion to highly polished surfaces is poor (Bowman and Meindl 1986). This coating can be patterned with masking tape, with an oxygen plasma (Loeb 1977) or with a special photolithographical procedure (UK Patent 1965). Although this parylene material looks interesting for the packaging of planar chemical sensors, it has not been used by the authors.

5.1.3.4 Printed circuit board strips.   As already indicated in figure 5.11, the sensor is mounted on a substrate. A low-cost package, suitable for laboratory experiments, has been built up around a printed circuit board strip. This method is used by virtually all research groups working on ISFETs or microelectrochemical sensors. It is simple and yet effective.

A single-sided printed circuit board (PCB) with the pattern shown in figure 5.13 has been fabricated using standard procedures. Copper is used as interconnection material and a commercial available connector is soldered onto the PCB. The PCB sensor strip is 9 mm wide and 90 mm long. The connector contacts are at the standard 2.54 mm distance.

The sensor is glued onto this substrate with non-conductive epoxy (Araldite). The interconnection between the aluminium bonding pads and the copper traces is made with Au bonding wires. The best adhesion to the copper traces is achieved with an ultrasonic ball bonder.

For the protection and encapsulation of the bonding wires, Epo-tek H54 epoxy is used in the bonding region. After hardening of this epoxy, Dow Corning Conformal Coating R-4-3117 is cast over the remaining parts of the strip. This silicone rubber is vulcanized at room temperature.

The operational lifetime of this encapsulation is more than two weeks. The encapsulation process, however, is completely manual, so this process is an expensive packaging method. The casting of the H54 epoxy and the R-4-3117 conformal coating is especially critical; the chemically active region has to be kept clear from these encapsulants.

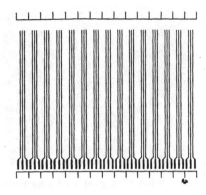

Figure 5.13   Copper pattern for the fabrication of printed circuit board strips.

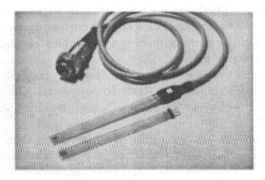

Figure 5.14   A packaged GLUCO sensor on a printed circuit board strip.

The printed circuit board strip has been used with success as a standard evaluation package for the voltammetric sensors reported in this book. A packaged GLUCO sensor is shown in figure 5.14. Increased performance of the package is obtained by using oxygen-plasma cleaning after the mounting of the sensor onto the PCB and before the wire bonding and the encapsulation. With this plasma cleaning, the surface contaminations are removed, resulting in an improved wire bond and better adhesion of the encapsulant material (Buckles 1987). This plasma cleaning also results in an improved electrochemical response of the sensor.

Problems were encountered with the manual application of the epoxy and the silicone rubber onto the bonding wires. It cannot be avoided that sometimes a drop falls on the chemically active part of the sensor. This results in a decrease in the active area of the sensor and therefore a decrease in the current. After a long (> 2 weeks) immersion in physiological solution, cracks are observed in the H54 epoxy layer. This deterioration of

Figure 5.15 Microphotograph of a corroded Al bond pad.

the epoxy is enhanced by high temperatures (> 40 °C). These cracks result
in the corrosion of the bonding wires and bonding pads. It is observed that
the sensor current first increases by a factor of between ten and a hundred
and then, after the complete corrosion, the current drops to very low levels.
The corrosion of the bonding pads can easily be seen under a microscope,
as indicated in figure 5.15.

After the same period of time as indicated for the H54 epoxy, the adhesion of the silicone rubber also starts to deteriorate. This problem can be
solved by the use of an appropriate adhesion promoter (e.g. Dow Corning
1204 primer).

It can be concluded that the printed circuit board strip can only be used
for the relatively short-term evaluation of planar electrochemical sensors.

5.1.3.5 Thick-film strips.   In the novel thick-film approach, the packaging
strip is fabricated with screen printing techniques. In figure 5.16, a photograph of a VOLTA I sensor mounted on a thick-film strip is shown. The size
of one strip is 50 mm by 5 mm, so that ten strips can be placed on one 2 × 2
inch substrate. The small width of the strip requires miniature connectors
with 1.27 mm between the contacts. Four masks and screen-printing steps
are necessary for the production of these strips.

1. During the first screen-printing step, Pd/Ag paste is used for the
interconnection of the bonding area and the connector soldering pad. This
paste has a good soldering ability.

2. The bonding area of the strip is screen printed in an Au paste with
a good bonding ability. This Au paste does not oxidize during storage of
the strip so that a reliable wire bond can be made to the bonding area.

3. The third layer is made in a chemically resistant dielectric paste.
The strip is completely covered with this paste, the bonding area and the
soldering pads excepted. This screen-printed layer replaces the silicone
rubber used in the printed circuit board approach.

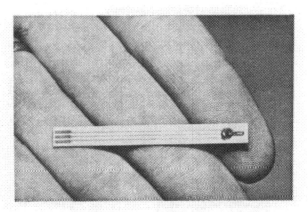

Figure 5.16   A packaged VOLTA sensor on a thick-film sensor strip

4. The last paste is a solder cream. This paste is applied on the soldering pads and allows the soldering of a surface-mounted connector.

After fabrication, the strips are scribed and separated by a numerically controlled laser. The sensors are glued onto the substrate with non-conductive Araldite epoxy. The interconnection between the aluminium bonding pads and the Au bonding area on the strip is made with Au bonding wires. For the protection and the encapsulation of the bonding wires, Epo-tek H54 epoxy is used. The application of silicone rubber is not necessary as the interconnection lines are protected by the dielectric layer.

Production automation and miniaturization are the major advantages of the thick-film strip in comparison with the printed circuit board approach. All steps in the production of this package can be done on a standard automatic production line for surface-mounted circuit boards, the application of the Epo-tek H54 epoxy excepted.

This critical step in the packaging can also be improved and automated with a method based on a re-usable stamp, described by Berg and Näbauer (1988). Before casting of the epoxy on the sensor, a stamp is pressed onto the sensor (figure 5.17). The form of the stamp is identical to the desired form of the opening in the encapsulation. After hardening of the epoxy, the stamp is removed mechanically. The stamp is made out of solid silicone rubber.

These stamps can be made on a mass-production basis using micromachining techniques in silicon. Firstly, the required shape of the opening in the encapsulation is etched in silicon with an anisotropic etchant. On this etched wafer, a silicone rubber is cast (White 1985). If necessary, the wafer can be coated beforehand with a release agent. After curing the silicone rubber, the stamps with the required form can be removed from the wafer. Preliminary results prove the feasibility of this technique. Optimum results have been obtained by the authors with Dow Corning Q3-3481

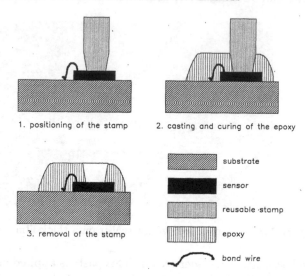

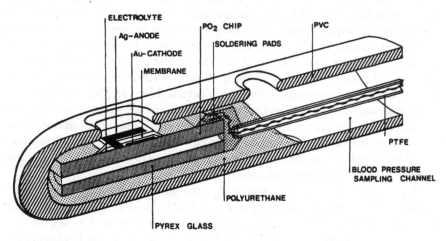

Figure 5.17   A re-usable stamp method for the packaging of chemical sensors.

Figure 5.18   A catheter-tip package for integrated sensors developed by Honeywell and Philips, Best, the Netherlands (now Dräger Medical Electronics) (from Engels and Kuypers (1983)).

mould-making material (Lambrechts 1989b).

5.1.3.6   Catheter-tip packaging.   The ultimate goal of planar sensors is the 'in vivo' measurement of biological parameters in the human body. Therefore, the sensor has to be packaged in a catheter. Catheter-tip packaging is more complicated than the packaging methods described above. As the dimension of the catheter has to be minimized, a mounting substrate in the catheter is not applicable. The long wires have to be connected directly to the sensor and their isolation has to be perfect. For short-term monitoring, a sample lumen is usually needed. Through this sample lumen, a blood sample can be withdrawn for calibration of the sensor. The most difficult problem remains the haemocompatibility of the assembly. Not only the choice of the materials, but also surface roughness and haemodynamic parameters are critical for the haemocompatibility. If the sensor assembly is not perfect, blood clotting can occur. The release of a blood clot into the circulatory system can then result in possible death of the patient. The risk of the 'in vivo' use of intra-vascular sensors has been described very clearly in Rolfe (1988). The responsibility of a manufacturer of 'in vivo' sensors is tremendous. Extensive animal experiments are necessary for the development of a reliable catheter-tip package.

A good example of how a sensor can be mounted into the tip of a catheter is shown in figure 5.18 (see also section 5.4.1). This figure is taken from Engels and Kuypers (1983).

5.1.3.7 A micromachined package. The ultimate goal of microelectrochemical sensor research is the realization of a smart sensor, hence microelectronic circuits have to be integrated with the sensor. These circuits have to be packaged very carefully, especially if they have to be implanted into the body. Blood is very corrosive and will destroy all circuits immediately. A Si_3N_4 passivation is a first step towards the protection of the electronic circuit. The protection of the bonding pads and overall protection is usually provided by epoxy; this works fine for short-term experiments, but fails in the long term.

A new idea based on micromachining techniques in silicon is proposed as a combined solution for the packaging of the interface electronics and the application of thick chemically active membranes. Micromachining has gained most interest in the world of chemical sensors, as this technique allows perfect packaging of the device. Several research groups are working in this direction (Blennemann *et al* 1987, van den Vlekkert *et al* 1987, Smith and Collins 1988), but most of them use glass as one part of their package. In our approach an integral silicon package is realized.

The proposed structure is assembled out of two silicon parts; a bottom part with interface electronics and an electrochemical sensor, and a micromachined top part with a micro-sieve and a cavity. The cavity can be filled (under vacuum) with a gel as an electrolyte or with a thick chemically active membrane. Both parts are hermetically sealed to each other. In this way an integral silicon package is realized (figure 5.19).

The bottom part consists of a CMOS interface circuit, as described in section 5.3.4, and a CMOS-compatible electrochemical sensor, such as the VOLTA II design for example (see further). The connection to the interface circuit is preferably made through the silicon wafer at the backside.

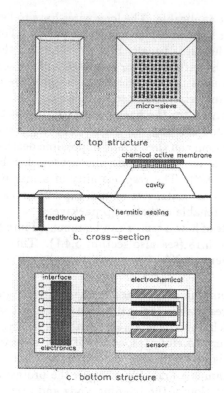

Figure 5.19   Schematic diagram of an integral silicon package for smart electrochemical sensors.

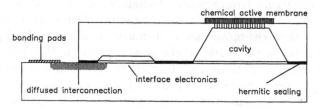

Figure 5.20   Adapted version of figure 5.19, where no feedthroughs are necessary.

This, however, implies that a cmos-compatible feedthrough has to be realized. The techniques reported in the literature, however, are either not cmos-compatible (Cline 1976, Erlich 1982, Kim 1985) or they need a large area per contact (van den Vlekkert 1988). Further research on a cmos-compatible feedthrough technique is therefore necessary. Meanwhile, the interface electronics have to be connected from the side by extending the bonding pads and shortening the top structure (figure 5.20).

Figure 5.21   SEM photograph of the micro-sieve.

The top structure has a shallow cavity to enclose the interface electronics and a large cavity that can be filled with, for example, an electrolyte or an enzyme membrane. Electrochemical connection of the cavity with the outer liquid is made via small holes etched into the silicon membrane. In this way a micro-sieve is formed. The cavities are etched in silicon with KOH.

Such a micro-sieve has been realized as a test structure for this new packaging technique (an SEM photograph is shown in figure 5.21). The outer dimensions of the cavity are 1.5 mm by 1.5 mm. The cavity is 300 μm deep and the silicon membrane is 50 μm thick; the diameter of the holes can be varied between 5 and 25 μm.

The package is completed by sealing the two parts to each other. Several techniques can be used for this purpose. The most elegant way is to sputter a thin Pyrex 7740 glass layer on the bottom of the micromachined structure and to seal both parts to each other using a low-voltage anodic bonding technique (Brooks and Donovan 1972). Possible alternatives are Au eutectic bonds and glass frit seals (Knecht 1987). Since the same grade of hermeticity for pressure sensors is not needed for electrochemical sensors, polymer sealing with screen-printed epoxies or photochemically patterned polyimide seem to be valuable alternatives.

If realized, this integral silicon package solves many problems. Since both parts are silicon, there is no thermal mismatch. The interface electronics are hermetically sealed from the corrosive environment; the influence of light on the interface circuit is also eliminated. Isolation of the bonding pads and power lines is easy, since the electrical connection is at the backside of the sensor. The fluidic, electrochemical connection is made on the other side. The cavity can be filled under vacuum with a gel or with a thick chemically active membrane. Since this membrane is mounted mechanically, adhesion of the membrane to the sensor is excellent.

For the development of this novel integral silicon package, however, a lot of work is still to be done on bonding or sealing techniques and on CMOS-compatible feedthroughs. It is nevertheless believed that only in this way

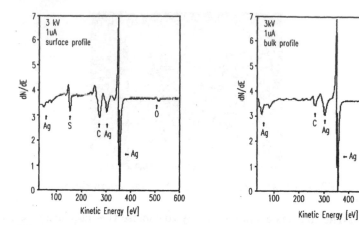

Figure 5.22   Auger spectrum of an evaporated (planar electrode) Ag film. Depth profiling reveals an essentially pure Ag layer.

can a durable long-term package be realized. Therefore, research on this interesting topic is essential.

5.2   ELECTROCHEMICAL EVALUATION OF THIN-FILM ELECTRODES

In the previous sections, a new technology for the fabrication of planar, CMOS-compatible, electrochemical sensors is described. A generic process based on polysilicon and a CVD oxide assisted lift-off method is presented. The different mask sets are detailed and the evolution from a crude prototype to a mature design is sketched. The fabricated devices have also been packaged in order to evaluate the different electrode materials. This evaluation is described in the next section.

Thin-film electrodes can be evaluated with different methods. The surface structure and purity of the sputtered or evaporated layer can be analysed with different techniques, such as scanning electron microscopy (SEM), Auger spectroscopy or secondary-ion mass spectroscopy (SIMS). All these methods have been used by the authors for the evaluation of the different layers. From the Auger spectrum of an evaporated Ag film it is concluded that this film is essentially pure Ag. As shown in figure 5.22, only a small amount of surface contamination (Ag_2S and C) is present on the sample. This contamination is due to the exposure of the sample to air and moisture and can, of course, influence the electrochemical response of the planar electrodes in an unpredictable way.

SIMS experiments on Ti/Pd/Ag multilayers also indicate that the top of this layer is pure Ag. During depth profiling, however, interdiffusion of the different materials into each other was suspected. This can be seen in

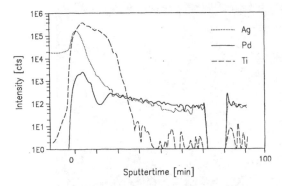

Figure 5.23 SIMS depth profile of a Ti/Pd/Ag multilayer.

figure 5.23. This 'interdiffusion' is, however, related to an artefact of the SIMS technique. It was observed that, during depth profiling, miniature Ag globules are formed on the surface, so a false image of the multilayer profile is measured with SIMS analysis. A more promising method for the analysis of Ti/Pd/Ag layers is thought to be Rutherford back-scattering RBS. Further investigation is therefore necessary. The influence of interdiffusion on the electrochemical response also remains unpredictable.

With SEM analysis, the surface structure can be determined. For example, when Ag is sputtered onto a contaminated surface, nodular growth of the Ag crystals is observed (figure 5.24). The surface structure will influence the electrochemical response of an electrode substantially; an increased surface roughness gives an increased active electrode area. For some applications this increased active area can be interesting, but usually a smooth layer and high electrode material purity are preferred.

It can be concluded from these initial remarks that it is useful to apply different analysis techniques to determine the quality of planar electrodes. The extrapolation of the spectroscopic results to the electrochemical response is, however, extremely difficult. Therefore, the most important method remains the characterization of the electrode with electrochemical techniques. It is evident that this approach gives the most valuable information, as these electrodes will be used in an electrochemical sensor.

In this book, the electrochemical response of a planar electrode is compared with the response of a classical bulk wire or ring electrode. A comparison with data from literature is also made, when available. As an electrochemical evaluation technique, cyclic voltammetry is used; this gives valuable information on electrode processes and data in the literature is usually available. As an electrolyte, a one molar H_2SO_4 solution is used. To avoid oxygen reduction currents, this solution is first saturated with nitrogen. Since the electrodes are intended for 'in vivo' use, cyclic voltammograms in physiological solution (PBS: phosphate buffer saline) have also

Figure 5.24   SEM microphotograph of a contaminated sputtered Ag layer showing nodular growth.

been recorded.

This approach has been used for the characterization of planar Pt, Pd, Au, Ag and C electrodes. All the electrode materials were applied on the sensor with the CVD oxide assisted lift-off technique, described earlier. As bulk electrodes, very pure noble-metal wires obtained from Johnson Matthey were used for comparison. A three-electrode potentiostatic measurement set-up was used with a standard Ingold Ag/AgCl reference electrode and a Pt Ingold ring auxiliary electrode.

5.2.1   Platinum electrodes

Platinum electrodes are used for a large variety of biosensors. They are used more specifically for biosensors that are based on the detection of H_2O_2 (e.g. glucose sensors) and for conductometric devices. Pt has the advantage that it is less reactive towards the Cl^- ions in the electrolyte under positive potential bias. Au and Ag electrodes cannot be used for this purpose as these electrodes will chloridate under these conditions.

A disadvantage of Pt electrodes is the extremely complex residual current. In figure 5.25, a typical cyclic voltammetric current–potential curve for a smooth platinum electrode in 0.5 M H_2SO_4 is shown. A great variety of peaks due to oxidation and reduction of adsorbed hydrogen and platinum oxide can be distinguished. These peaks are typical for a Pt electrode (see section 2.3.1). The residual current will interfere with the faradaic current due to the oxidation and reduction of the species to be sensed. This interference will be larger in the hydrogen region, since the peak currents in this

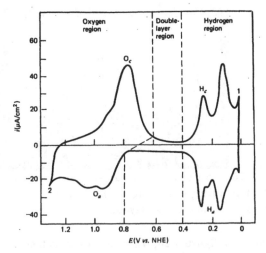

Figure 5.25 A typical cyclic voltammetric current–potential curve for a smooth platinum electrode in 0.5 M H_2SO_4 (from Bard and Faulkner (1980)).

region are not stable with time and environmental conditions. In the region of H_2O_2 detection (0.6–1.2 V versus NHE), the residual current is stable, allowing Pt electrodes to be used for this application. An alternative to Pt is the use of carbon electrodes. However, it is not obvious how to realize carbon electrodes using planar microelectronic techniques.

5.2.1.1 Preparation of the electrodes. The Pt layer is applied to the substrate by sputtering and is patterned with the CVD oxide assisted lift-off technique. The Pt is sputtered in a small table-top sputter machine made by Balzers (Med 010 Turbo). An optimization of the sputter process resulted in the following parameters:

base pressure: 2×10^{-6} mbar
substrate–target distance: 50 mm
sputter pressure: 0.03 mbar
sputter current: 50 mA
sputter rate: 25 nm min^{-1}
typical process time: 20 min
typical total thickness: 500 nm.

With these parameters, excessive heating of the substrate is restricted and a low-stress film is deposited. Destruction of the lift-off profile is therefore avoided. If the adhesion of the photoresist to the CVD oxide is insufficient, curling of the resist–Pt layer occurs even with these parameters. The adhesion between the resist–Pt film and the CVD oxide layer is then lost and the patterning of the Pt layer will be defective. An example of this

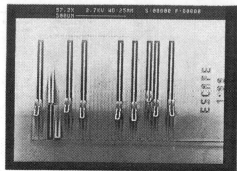

Figure 5.26   SEM picture of a catastrophic loss of adhesion of the resist–Pt film during sputtering of the Pt layer. This interdigitated structure is from the ESCAPE layout.

problem is shown in figure 5.26, where adhesion is lost on the interdigitated ESCAPE sensor structure. With an excellent resist adhesion, this problem can be avoided.

Adhesion of the sputtered Pt layer to polysilicon is sufficient for most applications. These layers survive the 'Scotch tape' adhesion test without any problem. However, under extreme conditions, such as a prolonged period in PBS or under high negative potential bias, the electrodes lose adhesion. To improve adhesion, a thin layer of Ti (50 nm) and Pd (50 nm) can be evaporated prior to the sputtering of the Pt layer. No direct influence of these thin adhesion layers has been seen by the authors in the electrochemical response of these planar Pt electrodes. However, in a recent publication (Josowicz *et al* 1988), some interference of TiW adhesion layers on the electrochemical response of their sputtered Pt electrodes is reported. The Pt layers used in that study are only 100 nm thick so that interference of the adhesion layer is more likely. It is claimed by Josowicz *et al* (1988) that Ti oxides block the active surface of the Pt layer and that these oxides can be removed with the proper electrochemical pre-treatment.

5.2.1.2   *Evaluation of sputtered Pt electrodes by cyclic voltammetry.* In figure 5.27, a cyclic voltammogram of a pure Pt wire in 1M H_2SO_4 is shown, as recorded with the measurement system described in section 3.2. If compared with figure 5.25 (a cyclic voltammogram from the literature), the peaks in the hydrogen region differ in a significant way. As indicated by Bard and Faulkner (1980) in figure 5.25, the shape, number and size of the peaks for adsorbed hydrogen depend on the crystal faces of platinum exposed, pre-treatment of the electrode, solution impurities and the supporting electrolyte. In this way, it is evident that deviations in the shape of the cyclic voltammogram in the hydrogen region can occur.

A cyclic voltammogram of a sputtered Pt electrode is shown in figure 5.28. The scaled voltammogram of figure 5.27 is also plotted on this

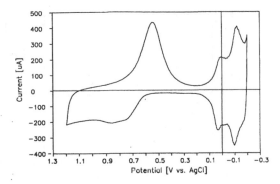

Figure 5.27 A cyclic voltammogram of a pure Pt wire in 1M H_2SO_4 saturated with N_2.

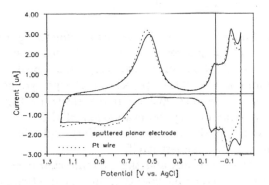

Figure 5.28 A cyclic voltammogram of a sputtered Pt electrode in 1M H_2SO_4 saturated with N_2.

figure for comparison. The voltammogram is scaled in such a way that the capacitive currents in the double-layer region match. A good correlation is found between the voltammogram of the sputtered electrode and the pure Pt wire. In the voltammogram, a third adsorbed hydrogen oxidation peak is observed. As discussed by Ross (1979), this can be related to a difference in the crystal orientation of the exposed crystal faces. Indeed, during sputtering, some preferential crystal growth can occur. For a bulk Pt wire, a random crystal orientation is assumed. Trace impurities on the metal surface or in the bulk metal can also induce these peaks.

On some wafers, this peak is even higher (figure 5.29). This indicates that the sputtering process has an important influence on the shape of the cyclic voltammogram in the hydrogen region. A similar response of a thin-film Pt electrode can be found in Koudelka *et al* (1987). However, as Pt electrodes are preferably not used in the hydrogen region, the exact shape of the hydrogen adsorption currents is not critical for the functioning of a H_2O_2-based glucose sensor.

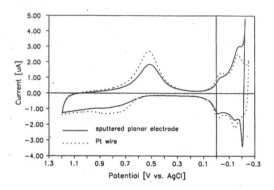

Figure 5.29 A cyclic voltammogram of a sputtered Pt electrode in 1M H_2SO_4 saturated with N_2. The adsorbed hydrogen oxidation peak is larger in this sample.

5.2.1.3 Contamination of the Pt electrodes due to Ag migration.   During the evaluation of the planar Pt electrodes in PBS, a strange phenomenon has been observed. Sometimes two very sharp peak currents appeared on the voltammogram. These peaks are always symmetrical around the zero potential of the Ag/AgCl reference electrode (figure 5.30). These peaks mainly occur during the evaluation of the complete planar three-electrode system. With an external auxiliary and reference electrode, this phenomenon did not manifest itself.

The shape of these peaks and the extreme sharpness indicate that this current is due to a reversible oxidation and a reduction of a contaminating species adsorbed on the surface of the working electrode. The location of the peaks around the zero potential suggest that this contaminating species is Ag. In other laboratories working on planar electrodes, similar peaks have been observed, but not reported. They related this behaviour to a contamination originating from the adhesion layers, such as Ti, Pd or Cr. The influence of Cr contamination on Pt electrodes was described in Koudelka *et al* (1987); the additional peaks observed in the voltammogram of Cr-contaminated Pt electrodes, however, does not correspond to the peaks observed in figure 5.30.

Next it is shown that the peaks depicted in figure 5.30 are due to Ag contamination of the working electrode. This contaminating Ag is deposited on the Pt working electrode during evaluation of the sensor due to Ag migration from the reference electrode to the working electrode. This assumption has been verified by the intentional contamination of Pt wire electrodes with the following metals: Ag, Cu, Pd and Ti. At first glance, none of these metals show analogous peaks, as seen in figure 5.30. The contamination of a Pt electrode with Ag, however, results in symmetrical peaks around the zero potential. In figure 5.31, the cyclic voltammogram of a Pt electrode with different grades of Ag contamination is shown. For a

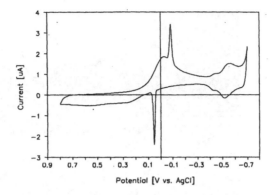

Figure 5.30   Cyclic voltammogram of an Ag-contaminated sputtered Pt electrode in PBS.

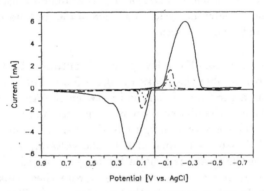

Figure 5.31   Cyclic voltammogram of a Pt wire electrode, intentionally contaminated with Ag, in PBS.

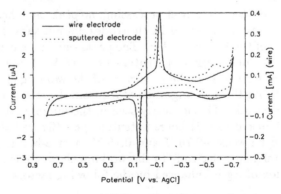

Figure 5.32   A comparison of the voltammogram of an intentionally contaminated Pt wire with the voltammogram of figure 5.30.

decreasing grade of contamination the height of the peaks and the spacing

Figure 5.33   Microphotograph of part of GLUCO sensor submitted to a 24 hours potential cycle. Migration of the Ag towards the working electrode is clearly seen.

between them decrease. For a low grade of contamination, the Ag-induced peaks start to resemble the peaks of figure 5.30. For a very low grade of contamination, a good match is found between the Ag-induced peaks in Pt wire electrodes and the observed phenomenon in the planar three-electrode configuration. This is shown in figure 5.32, where the voltammogram of the contaminated Pt wire is compared with the voltammogram of figure 5.30. The spacing of the peaks is a little bit larger; this can be related to the larger current at the Pt wire electrode. From these observations it can be concluded that the contaminating species is Ag. The process involved is the reversible oxidation and reduction of, respectively, Ag to AgCl and AgCl to Ag.

Initially, immediately after processing, no peaks are observed. The Auger analysis of freshly sputtered Pt electrodes did not show any trace of Ag at the surface of the Pt layer either. The peaks only appear when the sensor is submitted to high negative potentials (< -0.5 V). From these observation it can be concluded that the Pt surface of the working electrode is contaminated with Ag through migration of Ag from the reference electrode to the working electrode. This is clearly seen in the microphotograph of a GLUCO sensor submitted to a 24 hours potential cycle (figure 5.33). The potential was cycled between -0.75 V and 0.75 V. A tree-like migration structure from the reference electrode to the working electrode is seen.

Migration of Ag is enhanced by the following factors:

1. high negative potentials (< -0.5 V)
2. Cl^- ions in the solution
3. small WE–RE electrode spacing
4. non-uniform current distribution

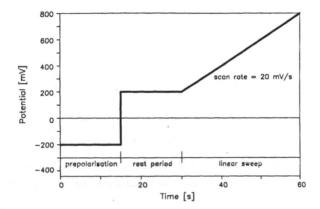

Figure 5.34 A typical waveform with a pre-polarization step for linear-sweep voltammetry at a Pt working electrode.

5. small working electrode area
6. large reference electrode area
7. excessive current flow through the reference electrode.

These factors should be avoided whenever possible. The GLUCO electrode configuration is, for instance, more sensitive to migration than the VOLTA I set. The VOLTA II set is also sensitive to migration due to the small working electrode area.

The best way to avoid migration is to use the sensor only in the positive potential range. No migration has been seen in planar Pt electrode configurations when used in the positive potential range. This corresponds to the range for which the planar Pt electrodes are intended; namely, the oxidation of H_2O_2 in biosensor applications.

5.2.1.4 Evaluation of the H_2O_2 sensitivity with LSV and amperometry. Since H_2O_2 is a product which is often generated in various kinds of biosensors, the response of Pt electrodes to H_2O_2 is of critical importance to the performance of the complete biosensor. The sensitivity of sputtered Pt electrodes to H_2O_2 has been investigated using two different measuring techniques, namely linear-sweep voltammetry and amperometry.

Linear-sweep voltammetry has been performed in PBS. A typical waveform with a pre-polarization step is shown in figure 5.34. The pre-polarization step is necessary for the activation of the Pt surface. Of importance are the scan rate, the amplitude and duration of the rest potential and the pre-polarization step. Good results in terms of residual current stability have been obtained with a rest potential of 0 mV for more then 15 seconds. A reasonable scan rate is 50 mV s^{-1}. The influence of the pre-polarization step was not clear. In Woods (1976) it is stated that the

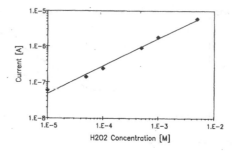

Figure 5.35   The response of a sputtered Pt electrode (VOLTA II) for different PBS–H_2O_2 concentrations using linear-sweep voltammetry.

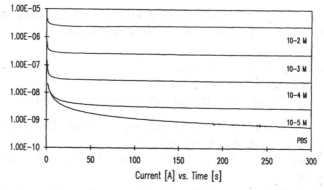

Figure 5.36   The current as a function of time for a sputtered Pt electrode for different PBS–H_2O_2 concentrations during an amperometric experiment.

electrodes can be maintained in the active condition by holding the potential in the oxygen adsorption region. A stable current is observed if the rest potential is less than 0 mV. At high rest potentials (100 and 200 mV), a continuous decrease in the current is observed. By introducing a pulse (e.g. +1000 mV) prior to the rest potential, the stability can sometimes be improved further. By applying a positive pulse (1000 mV, 5 seconds) the long-term stability of the residual current is ameliorated (Lambrechts 1989a).

A linear relationship with a good reproducibility was obtained (figure 5.35). This proves that planar Pt electrodes are suitable for H_2O_2 detection based biosensors.

In order to simplify the measuring equipment requirements, amperometric detection has also been investigated. The applied potential was kept constant at 700 mV. The current was then recorded as a function of time for five different H_2O_2 concentrations (figure 5.36). In this way, the significant range for H_2O_2 detection (10^{-5}–10^{-2} M) has been investigated. For

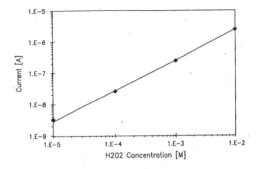

Figure 5.37 The response of a sputtered Pt electrode (VOLTA II) for different PBS–H_2O_2 concentrations during the amperometric experiment depicted in figure 5.36.

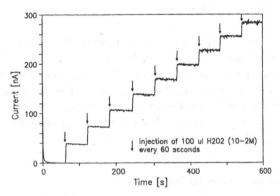

Figure 5.38 The current as a function of time for a sputtered Pt electrode for different stirred PBS–H_2O_2 concentrations during a continuous amperometric experiment.

amperometric measurements, a linear relation of the current as a function of H_2O_2 concentration was also found (figure 5.37).

In a second amperometric experiment, small amounts of a concentrated H_2O_2 solution were injected into the stirred measurement solution at well-defined time intervals. The current recorded during this continuous experiment is shown in figure 5.38. The resulting calibration curve is depicted in figure 5.39. An excellent linearity was obtained.

These experiments prove that the planar Pt electrodes described in this book can be used with success for H_2O_2 detection. It can be concluded that planar Pt electrodes behave identically in comparison with classical wire electrodes. A good match has been found between the cyclic voltammograms in H_2SO_4 and PBS of planar sputtered electrodes and wire electrodes. The sensitivity to H_2O_2 has been determined with linear-sweep voltammetry and amperometry and a linear response has been found. This

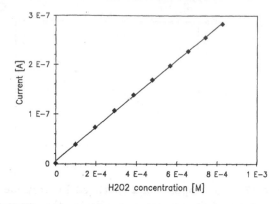

Figure 5.39   The response of a sputtered Pt electrode (VOLTA II) for different H_2O_2 concentrations during the amperometric experiment depicted in figure 5.38.

proves that the preparation method for planar Pt electrodes, based on the CVD oxide lift-off method, results in functional electrodes for H_2O_2-based biosensors. Under extreme experimental conditions, deviation from ideal condition was seen; small peaks appeared in the voltammogram. It is shown that these peaks are due to Ag migration from the reference electrode to the working electrode. Under normal operational conditions for H_2O_2 detection, these peaks do not appear.

5.2.2   Palladium electrodes

Palladium electrodes are rarely used in electrochemical sensor applications. It is, however, an interesting material because of the reactions with hydrogen. Palladium is capable of absorbing hydrogen atoms to a remarkable extent. Up to 0.69 hydrogen atoms are absorbed for each metal atom when a palladium electrode is in equilibrium with 1 atm hydrogen gas. This absorption results in a change in structure of the palladium metal from the α-phase to the β-phase when an atomic ratio of hydrogen to palladium higher than 0.05 is reached. This type of device (a palladium–hydrogen electrode) is used, for example, as a reference electrode in planar proton-conductor-based hydrogen sensors (Polak *et al* 1985). Palladium electrodes are also used for hydrogen detection as a Pd gate on ISFETs (Lundström 1981, Lundström and Sødderberg 1982).

As Pd is used as an intermediate adhesion layer for Ag, Au and Pt electrodes, Pd is also studied as an electrode material in order to determine the possible interferences. As a deposition method, electron-gun evaporation was used. The standard evaporation parameters, as used in a SLOAN evaporator, were: thickness= 500 nm, evaporation rate = 1.5 nm s^{-1}. A thin Ti adhesion layer was also evaporated prior to the Pd deposition

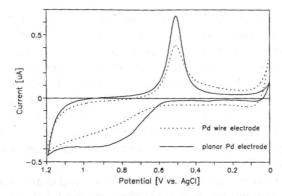

Figure 5.40   Cyclic voltammogram of a planar Pd electrode in comparison with a Pd wire electrode (1M H_2SO_4 saturated with nitrogen).

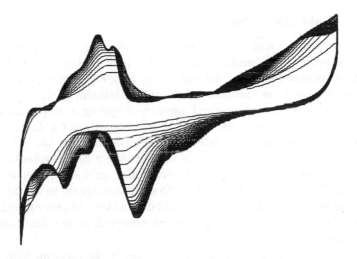

Figure 5.41   Twenty five consecutive sweeps of a Pt electrode with a thick Pd adhesion layer.

(thickness= 50 nm, evaporation rate = 2.5 nm s^{-1}). The palladium electrodes are patterned with the CVD oxide assisted lift-off technique.

In figure 5.40, the cyclic voltammogram of the evaporated, planar Pd electrode is compared with the voltammogram of a Pd wire electrode. A 1M H_2SO_4 solution saturated with nitrogen was used as electrolyte. Good agreement is found between the two curves. The shape of the curve also corresponds well with data found in the literature (Woods 1976).

In section 5.2.1 it is stated that the intermediate Pd adhesion layer does not have any influence on the electrochemical characteristics of the Pt electrodes. This is only true for thin (50 nm) adhesion layers. In figure 5.41, the influence of Pd on twenty five consecutive sweeps of a Pt electrode with a thick Pd intermediate adhesion layer is shown. A 500 nm thick Pd layer was placed erroneously between the 500 nm thick Pt layer and the 50 nm thick Ti layer. It is clear, when compared with figure 5.28, that such a thick adhesion layer has an important influence on the shape of the voltammogram; several extra peaks and a deformation of the existing peaks are seen. The voltammogram is a mixture of the voltammogram of Pd and the voltammogram of Pt. This corresponds well with information from literature where the electrochemical characteristics of a Pd/Pt alloy are studied (Woods 1976).

It can be concluded that the electrochemical characteristics of these planar Pd electrodes are identical to those of bulk Pd electrodes. Pd electrodes have not been studied extensively as the main use of these electrodes is in the field of electrochemical gas sensors.

5.2.3   Gold electrodes

Together with Pt and C electrodes, Au electrodes are frequently used in electrochemistry. Whereas Pt and C are used mainly for studies in the positive potential domain, Au electrodes may be advantageously applied in both the negative and positive potential domain. Gold, in contrast to Pt and Pd, adsorbs only a small amount of hydrogen at potentials before molecular hydrogen evolves. Gold is the noblest metal and only begins to adsorb oxygen at potentials more positive than for the other metals. Therefore, the only current that is flowing over a broad potential range is due to the charging of the double layer. Hence, gold electrodes are ideal for the detection of dissolved oxygen, as the voltammogram in the double-layer region is completely flat. For Pt electrodes, a large variety of peaks caused by reactions of adsorbed hydrogen are observed in the dissolved oxygen reduction region.

An important problem with Au electrodes is the reaction with Cl^- ions under positive potential bias. Due to this reaction, $AuCl_x$ is formed and the electrode material is consumed. Therefore, Au electrodes cannot be used as an auxiliary electrode in a three-electrode oxygen sensor; Pt electrodes are the only type that are suitable for this application.

Two different techniques have been used for the preparation of the Au electrodes. For the early GLUCO sensors intended for dissolved oxygen determination, a TiW/Au multilayer is used. The TiW alloy is sputtered as an adhesion layer (50 nm) prior to the resistive evaporation of the Au layer (500 nm) in the same vacuum cycle of an Alcatel sputter-evaporation system. For the more recent samples, a Ti/Pd/Au multilayer is used. The Ti layer (50 nm) and the Pd layer (50 nm) are evaporated onto the wafer in a multi-crucible electron-gun evaporation system using standard evaporation parameters. The Ti layer serves as an adhesion layer. The Pd layer

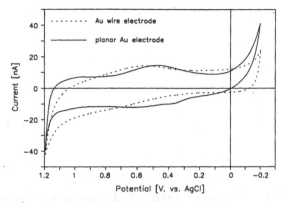

Figure 5.42 Cyclic voltammogram of a planar Au electrode in comparison with an Au wire electrode (1M H_2SO_4 saturated with nitrogen).

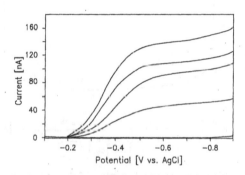

Figure 5.43 Linear-sweep voltammograms for an Au working electrode in PBS with different dissolved oxygen concentrations.

is necessary for the protection of the Ti layer against oxidation. After the evaporation of these two layers, the final Au layer is evaporated onto the wafer using resistive evaporation. Both metallization systems were patterned using the CVD oxide assisted lift-off method.

In figure 5.42, the cyclic voltammogram of the evaporated, planar Ti/Pd/Au electrode is compared with the voltammogram of an Au wire electrode. A 1M H_2SO_4 solution saturated with nitrogen was used as electrolyte. Good agreement is found between the two curves. If the shape of the curve is compared with data found in the literature (Woods 1976), a discrepancy is found. This is probably due to chloride ion contamination of the solution via the planar Ag/AgCl electrode. No difference was seen between the TiW/Au electrodes and the Ti/Pd/Au electrodes.

In figure 5.43, the linear-sweep voltammograms for an Au working electrode in PBS with different dissolved oxygen concentrations are shown. As discussed in section 2.3, the different regions in the voltammogram can be

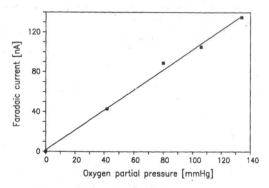

Figure 5.44 The oxygen response of a planar Au electrode; the current is plotted as a function of the pO_2.

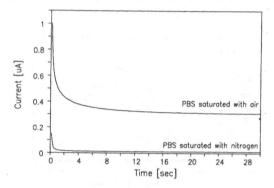

Figure 5.45 The current as a function of time for an amperometric measurement in PBS saturated with nitrogen and with air (i.e. the influence of oxygen).

clearly distinguished. In figure 5.44, the current is plotted as a function of the pO_2; a linear response is obtained. This proves that these planar Au electrodes can be used as the working electrode of an oxygen sensor.

To simplify the measurement, amperometry has been used as an evaluation tool. In figure 5.45, the current as a function of time is plotted for an amperometric measurement in PBS saturated with nitrogen and with air. For a zero pO_2 value only the capacitive current is recorded. When oxygen is present in the solution, this oxygen is reduced. During the first 30 seconds a declining current is observed; this current is described by the Cottrell equation (see section 2.3). After 30 seconds the current is approximately constant as a function of time. This current is also linearly proportional to the dissolved oxygen concentration and is a good measure of the pO_2 value.

During these measurements two unexpected phenomena occurred.

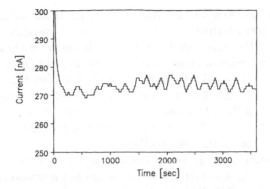

Figure 5.46 Low-frequency oscillations in the amperometric response of a planar Au electrode in PBS.

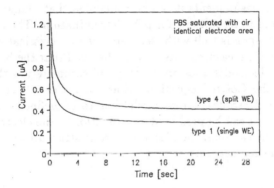

Figure 5.47 A comparison between the amperometric response of a type 3 and type 4 VOLTA I sensor in PBS saturated with air. Although the electrode area is identical, the current of the split electrode (type 4) is larger.

Firstly, it was observed that the current was not completely stable as a function of time; even ultra-low frequency oscillations (or 1/f noise) were observed (see figure 5.46). Secondly, the current was dependent on the electrode geometry. This was observed for the VOLTA I sensors; the current at the electrodes with a split active area (VOLTA I, type 4) was larger than for a normal electrode with the same area (see figure 5.47). This was due to an increased edge effect in the type 4 electrodes.

The stability of the Au electrodes is sufficient for most applications. In a detailed study of the stability of the VOLTA I and the VOLTA II Au electrodes (Van Hove and Placke 1988), it was found that for a six hour measurement, the response in PBS saturated with air varied only 6% for the VOLTA I and 3.5% for the VOLTA II electrodes.

These results are excellent, taking into account possible temperature and atmospheric pressure variations. Also, convection in the solution can occur, resulting in a current change. The existence of convection in the solution can also explain the difference between the electrode configurations. Since the VOLTA II electrodes are much smaller than the VOLTA I electrodes, they are less flow sensitive. A VOLTA II electrode has also been tested for its reproducibility as a function of time. The current was recorded daily for two weeks. During these fourteen days, a maximum relative deviation of 7% was recorded. As no provision was made for atmospheric pressure changes (> 20 mbar), these results are excellent.

The reproducibility of the sensors is not as good as expected. Electrodes from the same wafer and with the same electrode configuration can differ in response more than 20%. This implies that every sensor needs an individual calibration despite the high reproducibility of the electrode area obtained with the microelectronic fabrication techniques.

From these measurements it can be concluded that planar Au electrodes have been found to be suitable for pO_2 determination. The response of the electrodes corresponds well with the response of a classical wire electrode. For amperometric measurements, it was observed that the behaviour of the miniature Au electrodes does not correspond completely with the response predicted by the Cottrell equation. Since this Cottrell equation is only valid for large electrodes, a new model has to be defined. With this new model the current can be predicted as a function of the electrode geometry. It has to take into account the planar electrode structure and the extremely small electrode geometries.

5.2.4   Lifetime and stability of Ag/AgCl electrodes

As each electrochemical sensor needs a reference electrode to define an accurate electrochemical potential in the electrolyte solution, the behaviour of these reference electrodes is of critical importance for a reliable response of the entire sensor. Silver/silver chloride electrodes are therefore of special interest for the development of biosensors. The potential at this electrode depends on the Cl^- ion concentration in the electrolyte. As the Cl^- ion concentration in blood is almost constant, the potential at a Ag/AgCl electrode in blood does not change substantially. However, blood is very corrosive, so the corrosion resistance of the planar Ag/AgCl electrodes is of prime importance. Other important parameters for reference electrodes are the drift, stability, reproducibility and the noise behaviour. All these different aspects have been investigated for planar Ag/AgCl electrodes prepared with the CVD oxide assisted lift-off technique.

5.2.4.1   Preparation of planar Ag/AgCl electrodes.   Silver thin-film electrodes have been prepared with the CVD oxide assisted lift-off technique

Table 5.2   Evaporation parameters of the different layers.

Metal	Thickness [nm]	Evaporation rate [nm/sec]
Ti	50	2.5
Cr	50	2.0
Pd	50	1.5
Ag	900	10.0

using polysilicon as interconnection material. As is widely known, adhesion of silver thin films to silicon or silicon dioxide is poor. Usually, an intermediate layer of Ti or Cr is used to promote adhesion. However, these adhesion layers corrode easily: mechanical and electrical contact is then lost. Therefore, loss of adhesion is one of the main failure mechanisms of planar Ag/AgCl electrodes. Four different metallization methods have been examined by the authors: Ag, Ti/Ag, Cr/Ag and Ti/Pd/Ag multilayers. The metals were deposited sequentially with an electron-gun evaporation system. The deposition parameters are summarized in table 5.2.

Different classical methods exist for the conversion of Ag electrodes into Ag/AgCl electrodes; this conversion is named 'chloridation'. The chloridation of the silver electrodes can be done electrochemically in an NaCl or HCl solution by applying a voltage of 0.7 V, or chemically by dipping in an $FeCl_3$ or $KCrO_3Cl$ solution (Shumilova and Zhutaeva 1978).

An alternative for the fabrication of Ag/AgCl electrodes is the evaporation of AgCl on an Ag layer. A comparison was made between the chemical chloridation based on $KCrO_3Cl$ and the evaporation of AgCl (Bousse *et al* 1986). It was concluded that the chemical approach has several advantages. The electrodes are more stable, have a lower series impedance and are simpler to fabricate.

A chemical chloridation method based on $FeCl_3$ has been developed by the authors. This method is frequently used for the chloridation of encapsulated electrodes. The principle, however, can be converted to a mass-production technique using standard microelectronic procedures. Standard photoresist can be used for the protection of the aluminium pattern and for the selective chloridation. The protection of the aluminium pattern is necessary as aluminium is heavily attacked by $FeCl_3$. In figure 5.48, a microphotograph of a Ag layer that has been selectively chloridated using photoresist as masking material is shown. The black areas on the photograph are Ag/AgCl; the white areas are pure Ag.

It was experienced that the most efficient way to chloridate a planar electrode consists of the chloridation of the complete wafer immediately

Figure 5.48   Microphotograph of an Ag layer, selectively chloridated using photoresist as a masking material (resolution marks are 10 μm wide).

after the deposition of the Ag layer and before the lift-off of this Ag layer. The optimum procedure for the fabrication of CMOS-compatible Ag/AgCl electrodes can then be summarized as follows:

1. standard CMOS process;
2. preparation of the CVD oxide assisted lift-off profile;
3. evaporation of a Ti/Pd/Ag multilayer (50/50/900 nm);
4. dipping of the wafer in deionized water in order to wet the Ag layer (This wetting improves the homogeneity of the AgCl formation);
5. chloridation of the Ag in a 1% FeCl$_3$ solution by dipping for one minute;
6. lift-off of the Ti/Pd/Ag/AgCl multilayer.

The FeCl$_3$ concentration and the chloridation time were optimized with the thickness and the uniformity of the AgCl layer as criteria. A one minute dip in a 1% FeCl$_3$ solution results in a uniform AgCl layer. The thickness of this layer was rated at 0.6 μm with an Ag consumption of 0.2 μm. The total multilayer structure is then: 50 nm Ti, 50 nm Pd, 800 nm Ag, 600 nm AgCl. As seen on the SEM picture (figure 5.49), the surface structure of this AgCl layer is granular with pores. The diameter of the grains and pores is approximately 200 nm. As suggested in Bousse *et al* (1986), the porosity of a AgCl layer can have a large influence on the impedance of the electrodes. This impedance aspect of planar Ag/AgCl electrodes has not been investigated in this study.

The potential of the different Ag/AgCl electrodes has been measured against a conventional Ingold Ag/AgCl reference electrode with an internal reference electrolyte. The Cl$^-$ concentration in the solution was varied between 10^{-4} M and 1 M. Initially the behaviour of these electrodes is Nernstian for all metallization methods. A slope of 58 mV/pCl$^-$ was observed (see figure 5.50). However, the Ag, Ti/Ag and Cr/Ag electrodes started drifting after a few hours; this was related to loss of adhesion of

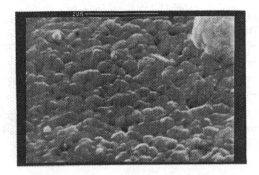

Figure 5.49 SEM photograph of the surface structure of an Ag/AgCl layer.

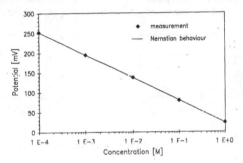

Figure 5.50 Nernstian response of a planar Ti/Pd/Ag/AgCl electrode to a varying [Cl$^-$] concentration.

the metal layer. No significant drift was observed with the Ti/Pd/Ag electrodes. To examine this corrosion phenomenon, an accelerated lifetime test has been set up.

5.2.4.2 Corrosion resistance. Twenty four electrodes per metallization were immersed in a physiological solution at room temperature for three months. Periodically the solution was ultrasonically agitated to remove the corroded parts of the metallization. With the pure silver electrodes, adhesion was lost immediately (figure 5.51(a)). Loss of adhesion of the Ti/Ag was observed after a few hours (figure 5.51(b)). The average lifetime was 2.75 hours. For both metallization methods, the complete electrode lost adhesion. Corrosion of the Cr/Ag electrodes, in contrast, started with the formation of blisters and holes. The average time before blister or hole formation was 46 hours (figure 5.51(c)).

The Ti/Pd/Ag electrodes, however, withstand this lifetime test without any problem. After three months, no corrosion or loss of adhesion was observed (figure 5.51(d)). For chloridated Ag, Ti/Ag and Cr/Ag samples, the average lifetime was halved. No damage was found on the chloridated

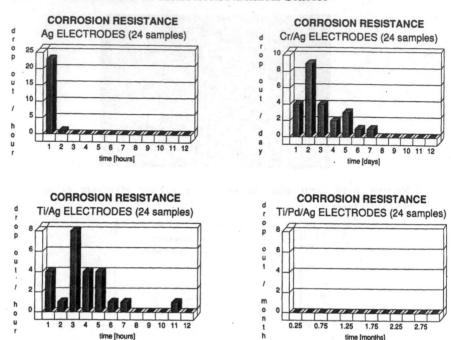

Figure 5.51   Lifetime profiles of the different metallization methods.

Ti/Pd/Ag electrodes after 3 months. This proves that Ti/Pd/Ag electrodes are far superior to all the other metallization methods.

Another research group obtained similar results for solar-cell metallizations (Sharp 1979). They related the excellent corrosion resistance to a modification of the anodic potential of the active metal by the introduction of palladium, so that a passive oxide is formed when the film is exposed to an electrolyte. Palladium also serves as diffusion barrier for oxygen and moisture. It inhibits the corrosion of the Ti layer and so preserves adhesion of the Ag film. The results of this corrosion study have been published in Lambrechts *et al* (1986a,b). Since then several research groups have adopted this method. Their Ag/AgCl reference electrodes, based on a Ti/Ag multilayer, also suffered from corrosion problems.

5.2.4.3   Noise and stability.   Ti/Pd/Ag/AgCl electrodes have been used with success as reference electrodes in the different voltammetric sensors developed by the authors. For this application, the stability of the reference electrode is not so critical, since the current of a voltammetric sensor is measured on a current plateau; small deviations of the reference potential (±5 mV) will not influence the response of a voltammetric sensor.

For potentiometric sensors, however, these deviations are inadmissible.

For a pH electrode, a variation of 5 mV at the reference electrode corresponds to a variation of 0.1 pH. Since for biomedical applications the pH value has to be measured with an accuracy better then 0.02 pH, the accuracy and stability of the reference electrode has to be better than 1 mV. This implies that a reference electrolyte has to be used and that the stability of the Ag/AgCl electrode itself has to be excellent.

The long-term stability of an Ag/AgCl electrode is difficult to measure if the measurement period is extended to more than 24 hours. The most elementary way is to measure the drift of the potential between a commercial reference electrode and the planar Ag/AgCl electrode in an electrolyte with a constant Cl$^-$ ion concentration. This potential has to be measured with a high-accuracy, high-impedance voltmeter. Although this straightforward method looks easy, it is difficult to implement this technique in reality for long-term measurements. One has to take into account the drift of the commercial reference electrode, the temperature variations and the Cl$^-$ concentration variations due to evaporation of the liquid. These problems can be solved by hermetically sealing the electrolyte from the environment and by the use of a high-quality reference electrode and a thermostatic bath. The main problem, however, remains the leaking of the reference electrolyte out of the reference electrode. This leakage of reference electrolyte cannot be avoided, as a liquid contact is essential for the correct functioning of the reference electrode.

An alternative is to measure the potential between two identical planar Ag/AgCl electrodes in a sealed electrolyte under thermostatic control. With this method the leakage of reference electrolyte is avoided. However, systematic drift cannot be measured as under identical conditions both identical electrodes will drift in an identical way. This can be avoided by the use of a good Ag/AgCl wire electrode as reference electrode. However, the problem remains that the combination of the drift of the conventional electrode and the planar electrode are still measured together. It can be concluded that the measurement of the long-term stability remains a problem.

Proof that our planar Ti/Pd/Ag/AgCl electrodes are very stable is the observation that the potential difference between an electrode which is more than two years old and a newly fabricated electrode is usually less then 2 mV. Since the current of a voltammetric sensor is always measured at a current plateau, this small deviation in potential will not influence the voltammetric sensor response at all. It was also seen that the short-term stability increases if the chloridation of the Ag layer is performed immediately after the processing of the Ag electrode. After prolonged exposure of an Ag layer to the atmosphere, contamination is observed. This contamination consists mainly of Ag_2S and decreases the short-term stability of the sensor. Because of the surface contamination, a longer aging procedure is necessary. Hence, it is preferable to chloridate the Ag electrodes on the wafer immediately after the evaporation of the Ag layer and before lift-off of the metal.

As an alternative method, stability of thin-film Ag/AgCl electrodes has been correlated to the noise characteristics of the electrode interface. The results obtained with this method have been published in Steyaert *et al* (1987). The noise at an electrochemical interface is generated by two sources; the electrode interface and the bulk electrolyte resistance. From measurements of the noise level as a function of the electrode distance, it is concluded that the contribution of the bulk electrolyte resistance can be neglected in comparison with the noise generated at the electrochemical interface. The noise at the electrode–electrolyte interface has been fitted to the following relation:

$$Sv(f) = \underbrace{\frac{4kTK_1}{Af[\text{Cl}^-]}}_{1/f \text{ noise}} + \underbrace{\frac{4kTK_2P}{A^{0.5}[\text{Cl}^-]}}_{\text{white noise}} \tag{5.1}$$

where $Sv(f)$ is the equivalent voltage noise spectral density ($\text{V}^2 \text{ Hz}^{-1}$),
K_1 and K_2 are the constants,
k is the Boltzmann constant (J K^{-1}),
T is the absolute temperature (K),
A is the electrode area (mm^2),
P is the electrode perimeter (mm),
$[\text{Cl}^-]$ is the chloride ion concentration (M).

Two different parts of equation (5.1) can be distinguished: the $1/f$ noise and the white noise. The validity of this relation has been verified by noise measurements on different electrode configurations. For more details on the measurement configuration and experimental results the reader is referred to Steyaert *et al* (1987).

It is clear from equation (5.1), that a low form factor ($P/\sqrt{A}$) results in a low white-noise level. So, the noise generated by a circular electrode is lower than that generated by a square electrode with the same area. Rectangular electrodes generate even more noise. From this point of view, the electrode configuration used in the VOLTA layout series should be avoided. In practice, however, the geometrical constraints of an implantable sensor lead to such an electrode configuration. Fortunately, the noise levels are still acceptable for most applications.

Also important to notice is the influence of electrode area on $1/f$ noise. The $1/f$ noise is inversely proportional to the electrode area. This implies that small electrodes are less stable than large electrodes, so $1/f$ noise limits the scaling of electrochemical sensors to extreme small dimensions.

In figure 5.52, the voltage noise-power spectrum is plotted for different electrode materials for an identical electrode configuration. From these measurements it can be concluded that the white-noise levels are identical.

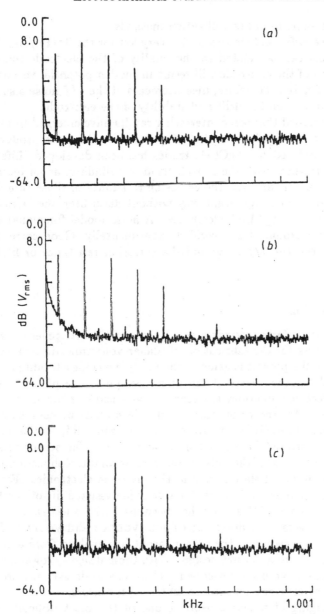

Figure 5.52 Voltage noise-power spectrum for the different electrode materials: (*a*) Ti/Ag/AgCl, (*b*) Cr/Ag/AgCl, (*c*) Ti/Pd/Ag/AgCl.

This is related to the fact that the white noise is generated at the electrode–electrolyte interface. At this interface the electrode material is Ag/AgCl and is thus the same for all metallization methods.

The $1/f$ noise differs extensively and is very low for the Ti/Pd/Ag/AgCl electrodes. This can be related to the quality of the electrode material itself. Corrosion of the electrode will result in sudden potential variations, resulting in $1/f$ noise. Therefore, measurements of the $1/f$ noise also give valuable information on the drift and stability of the electrode.

It can be concluded that some interesting results have emerged from this study on planar Ag/AgCl electrodes. A process for the mass production of on-wafer chloridated Ag/AgCl electrodes has been developed. Lifetime tests have shown that, with the introduction of palladium as an intermediate layer between Ti and Ag, the corrosion resistance of silver thin films has been improved in a significant way without disturbing the Nernstian characteristics of the Ag/AgCl electrodes. A noise model for planar electrodes has been introduced and verified experimentally. Good correlation was found between the $1/f$ noise and the corrosion resistance or lifetime of the electrodes.

5.2.5   Conclusion

In this section the efforts towards a novel, state-of-the-art, generic, CMOS-compatible process for the fabrication of planar voltammetric sensors are described. Since the process is generic, the newly developed techniques can also be applied to other electrochemical sensor types, like planar oxygen sensors, ion-selective potentiometric sensors, gas sensors or conductometric cells. The basis of the sensor process is a standard CMOS process, so CMOS interface circuits can easily be integrated onto the same chip as the sensor, as will be demonstrated in sections 5.3.1 and 5.3.4. The sensor process itself is based on the use of the polysilicon layer as an interconnection layer between the noble-metal electrodes and the interface electronics. For the patterning of the noble-metal layers, the CVD oxide assisted lift-off method, developed in the scope of this book, has been used with success.

Different mask sets were designed; GLUCO, VOLTA I and VOLTA II. During the years of our research, the sensor layout has evolved from a crude, imperfect prototype to a mature design. With these designs, voltammetric electrode systems have been developed. These electrode systems are the essential detector elements for glucose and oxygen sensors.

As the packaging of planar sensors is one of the most difficult tasks in the development of electrochemical transducers, considerable attention has been paid to this topic. Several innovative packaging techniques have been developed, based on printed circuit board or thick-film sensor strips. A great variety of epoxies and silicone rubbers have been tested for this purpose: Epo-tek H54 and Dow Corning Conformal Coating R-4-3117 are good encapsulants for electrochemical sensors.

Packaging is necessary for the electrochemical evaluation of the thin-film electrodes. Planar Pt, Pd, Au, Ag and C electrodes have been prepared using the CMOS-compatible sensor process. These electrodes have been evaluated with electrochemical measurement techniques and compared with pure bulk wire electrodes. It is concluded that thin planar electrodes behave identically to conventional metal electrodes. This proves that reliable electrochemical sensors can be made with the technology described.

Since each electrochemical sensor needs a reference electrode to define an accurate electrochemical potential in the electrolyte solution, the behaviour of Ag/AgCl reference electrodes has been investigated extensively. Lifetime tests have shown that with the introduction of palladium as an intermediate layer between Ti and Ag, the corrosion resistance of silver thin films has been improved in a significant way without disturbing the Nernstian characteristics of the Ag/AgCl electrodes. A noise model for planar electrodes has been introduced and verified experimentally. Good correlation was found between the $1/f$ noise and the corrosion resistance or lifetime of the electrodes. It can be concluded that with the tools developed in this chapter, a reliable detector element for glucose and oxygen sensors has been realized. Pt electrodes can be used for H_2O_2 detection based biosensors. Au electrodes can be used for pO_2 measurements. The next step in the development of glucose and oxygen sensors is the preparation of the planar membranes on top of the voltammetric electrode systems described.

5.3 PRACTICAL REALIZATIONS OF MICRO-ELECTROCHEMICAL BIOSENSORS AT K U LEUVEN

The major applications of the microelectrochemical devices described in this book are as O_2 or H_2O_2 detectors for biosensors. However, the technology developed is generic, so potentiometric ion-selective electrodes and conductometric biosensors can also be realized.

In this section, smart planar ion-selective electrodes, planar voltammetric glucose and oxygen sensors, a smart voltammetric and a planar conductometric urea sensor are detailed. All these sensors are developed by the authors and their co-workers in ESAT–MICAS. These sensors could not have been realized without the help of J Suls, A Claes, P Jacobs, M Steyaert, K Vriens, D De Wachter and L Callewaert.

5.3.1 Planar ion-selective electrodes

Interesting smart electrochemical sensors can also be made with ion-selective electrodes. An excellent example is the revolutionary eight-channel ion-selective sensor developed by our group in a joint effort with

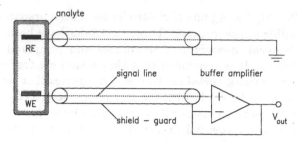

Figure 5.53 Schematic diagram of an integrated ion-selective electrode according to the EGFET approach.

Jan Van der Spiegel (Moore School of Electrical Engineering, University of Pennsylvania, Philadelphia, PA, USA) and Imants Lauks (Integrated Ionics Inc, Princeton, NJ, USA). This smart sensor is based on the concept of the extended-gate FET (EGFET) as described in van der Spiegel *et al* (1983), but instead of a source follower a complete buffer amplifier is used (Lauks *et al* 1985). The design of the masks for the chemically active areas was a contribution of the authors. The operational amplifier was designed by Michiel Steyaert (Steyaert 1987).

In figure 5.53, the schematic diagram of one integrated ion-selective electrode line is shown. In contrast with an ISFET, the electrochemically sensitive site is spatially separated from the transistor circuits to facilitate the packaging of the interface electronics. This implies that an EGFET is a genuine potential detector, whereas an ISFET is more of a charge detector.

An ion-selective electrode is connected with an operational amplifier in unity-gain follower mode by a polysilicon signal line, hence the name EGFET. This signal line is surrounded by an insulated shield, providing electrical as well as chemical protection of the signal line and reducing cross-talk between the different sensors. By bootstrapping the shield, the capacitive coupling between the signal line and the shield is reduced. Resistive leakage currents from the signal line to areas with a different potential are also minimized. This on-chip coaxial transmission line was made in a 5 μm CMOS three-poly process (Callewaert and Sansen 1984). The signal line was made out of the middle poly layer; the top layer and bottom layer formed a shield by interconnecting these layers.

In figure 5.54, the mask layout of the eight-channel ion-selective sensor interface is shown. The concept of spatial separation of electronics and electrochemistry can be clearly seen. Eight different sensors with their corresponding buffer amplifiers and one reference electrode are placed on the device. The chip measures 2.8 mm by 3.3 mm.

For the realization of one ion-selective electrode, up to three extra sensor layers are necessary. The functions of the different sensors are summarized in table 5.3 and the subsequent layers are indicated. Ion-selective elec-

SENSOR M.L.

MICAS M.S.

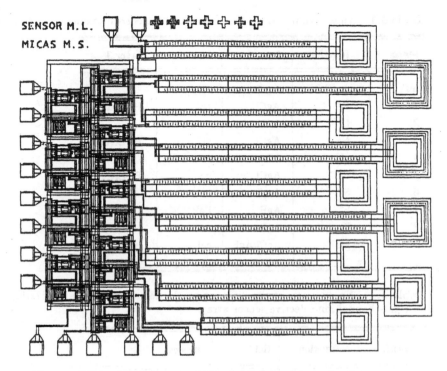

Figure 5.54   Mask layout of the eight-channel ion-selective sensor interface.

trodes are provided for K^+, Cl^-, Na^+, pH, Ca^{2+}, S^{2-}, F^+ and Br^-. In the design, provision is also made for changing two ISEs into a gas sensor. For this eight-channel realization, twelve extra layers and masks are necessary. The function and the field of these masks are indicated in table 5.4, in combination with the preferential processing technique for the corresponding layer. A microphotograph of the sensor chip prior to sensor processing is depicted in figure 5.55.

From this list, it is clear that processing of such a device is very diffi- cult. Without a gas sensor, ten different non-standard photolithographic processing steps are necessary. All these layers are also chemically sensitive and can easily be contaminated or destroyed by maltreatment. This im- plies, for example, that the processing of the tenth layer must not influence or change the characteristics of the nine previous layers. Since the realiza- tion of a single planar ion-selective sensor is already quite an achievement, it is clear that this eight-channel ion-selective sensor is more science fiction than reality. This design, however, indicates nicely what will be possible in the field of planar sensors in the next five years if a commercial background is found for these applications.

As a test case, a more realistic device has been made with this struc-

Table 5.3   An overview of the different ISEs and the corresponding construction.

sensor	function	layer 1	layer 2	layer 3
ref	AgCl RE	AgCl	-	-
1	K^+	AgCl	gel	PVC(K^+)
2	Cl^-	AgCl	-	-
3	Na^+	AgCl	gel	PVC(Na^+)
4	pH	IrO_2	-	-
5	Ca^{2+}	AgCl	gel	PVC(Ca^{2+})
6	S^{2-}	Ag_2S	-	-
7	F^+	AgF	LaF_3	-
8	Br^-	AgBr	-	-
5+7	gas	AgCl/AgF	gel	teflon

Table 5.4   An overview of the different masks necessary for the realization of an eight-channel ion-selective sensor array.

mask	function	field	process
M1	IrO_2	dark	lift-off
M2	Ag_2S	dark	lift-off
M3	AgBr	dark	lift-off
M4	AgCl	dark	lift-off
M5	AgF	dark	lift-off
M6	LaF_3	dark	lift-off
M7	gel	light	photolithography
M8	PVC(K^+)	light	dry etching
M9	PVC(Na^+)	light	dry etching
M10	PVC(Ca^{2+})	light	dry etching
M11	gel	light	photolithography
M12	teflon	light	dry etching

ture. Eight planar Ag/AgCl pseudo-reference electrodes have been made with the CVD oxide assisted lift-off method described in chapter 4. The response of five identical electrodes to a varying Cl^- concentration is shown in figure 5.56. The offset between the different electrodes is due to uncompensated offset of the operational amplifiers, not to offset of the individual Ag/AgCl electrodes. The offset problem has been corrected in a second version of the sensor interface. This measurement clearly indicates that the

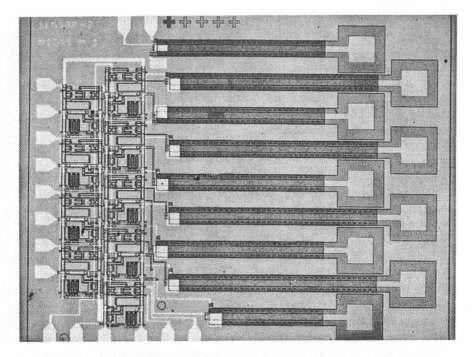

Figure 5.55 Microphotograph of the eight-channel sensor chip prior to sensor processing.

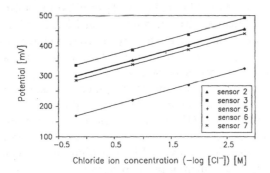

Figure 5.56 Eight-channel ion-selective sensor Response of five Ag/AgCl electrodes to a varying chloride ion concentration.

principle of the EGFET works in practice and that ion-selective electrodes can be combined into a smart sensor.

It can be concluded that with this eight-channel ion-selective sensor array, a completely new generation of planar chemical sensors has been initiated. The commercial success, however, depends on the economic viability of this very complex device; enormous numbers of this sensor have to be produced and marketed. This can only be done if the main advantage of such a sensor array is exploited; by a complete characterization of this device, a complete sensitivity matrix of all the sensors for all the ions can be set up. All cross-sensitivities of one ion-selective electrode to the other ions can then be eliminated by solving eight equations with eight unknowns. This requires, however, that all the elements of the sensitivity matrix remain constant and are not influenced by external parameters such as temperature or sensor history.

5.3.2 Glucose sensors

In this, and in the following section, special attention is paid to glucose sensors and dissolved-oxygen sensors, since there is an urgent need for these sensors in the medical world. The most interesting application is the long-term 'in vivo' use of a glucose sensor for an artificial pancreas. However, an operational lifetime of more than one year is an essential requirement for this application. This requirement cannot be realized with the present status of GOD membrane technology. Problems are biocompatibility and the decay of enzyme activity.

Current research is therefore oriented towards short-term 'in vitro' or 'in vivo' measurements with disposable sensors. A disposable glucose sensor with an operational lifetime of 48 hours would be very useful for monitoring a new diabetic patient, so that the functionality of the patient's glucose metabolism could be measured. Single-use, cheap disposable sensors can also be useful for the daily blood glucose determination of diabetic patients. Dissolved-oxygen sensors with an operational lifetime of a few hours are needed for 'in vivo' monitoring of the oxygen concentration in blood during anaesthesia and surgery. Sensors with a longer lifetime can be used for monitoring patients in intensive care units.

The principle of a GOD enzyme based glucose sensor is detailed in section 2.3.2. In this book, a planar GOD enzyme glucose sensor with a H_2O_2 detector is detailed. A GOD–BSA membrane is therefore applied with lift-off techniques on top of a CMOS-compatible electrode structure. To cope with the oxygen deficiency problem at high glucose concentrations, a dry-etched polyurethane membrane is proposed.

5.3.2.1 Preparation. Although membranes have been applied with dip-coating techniques on mounted and packaged sensors, in this section only on-wafer techniques are discussed. All steps necessary to prepare a planar sensor have already been discussed separately in this book, so the following procedure for the preparation of a planar glucose sensor is proposed.

1. Processing of the CMOS-compatible electrode structure (section 5.1).
For the H_2O_2 detector element, a planar CMOS-compatible three-electrode configuration is used. The working and auxiliary electrodes are

both made in Ti/Pd/Pt. As a reference electrode, Ti/Pd/Ag/AgCl is used. The 'GLUCO', 'VOLTA I' and 'VOLTA II' mask set are all suitable for this purpose. The optimum electrode configuration is the 'VOLTA II' structure.

2. Lift-off of the GOD–BSA membrane (section 4.4.4).

The enzyme glucose oxidase is cross-linked into a bovine serum albumin matrix. For good adhesion, a surface pre-treatment (detailed previously) is necessary. As this method is based on surface hydroxyl groups, the adhesion to the electrodes is not always satisfactory. Some solutions to this problem will be proposed further.

3. Dry etching of the polyurethane membrane (section 4.4.5).

Although this method has not yet been evaluated completely, it is believed that it will result in a working glucose sensor with extended linear range.

4. Wafer dicing and packaging of the sensor (section 5.1.3).

The only process step that cannot be carried out on wafer level is the packaging of the individual sensor. In order to obtain the low production cost imperative to satisfy a market for disposable, single-use glucose sensors, packaging also needs automation.

The optimization of this process has been carried out by one of our co-workers, Annick Claes, and resulted in a working planar CMOS-compatible glucose sensor.

5.3.2.2 Results.   In figure 5.57, a microphotograph of a GOD–BSA membrane on a 'VOLTA I' electrode structure is depicted. An excellent pattern delineation can be observed. The amperometric response of this planar CMOS-compatible glucose sensor is shown in figure 5.58. A linear range up to 10 mg dl^{-1} (180 mM) is obtained without a diffusion membrane. Since this response is one of the first measurements on a complete planar glucose sensor, further optimization and evaluations are still necessary for the complete characterization of this device.

This and other similar measurements prove that, with the techniques described in this book, a working planar CMOS-compatible sensor can be produced. Prior to the 'in vivo' use with diabetic patients, some critical problems have to be solved, however.

5.3.2.3 Problems.   Although all techniques were made available for the realization of a planar glucose sensor, some problems were encountered during the combination of the different process steps.

Because of the high temperature during the curing of the Epo–Tek H54 epoxy (80–100 °C), this material cannot be used for the encapsulation of the bonding wires. Therefore, a new epoxy has to be found for the development of a low-temperature package. After evaluation of different epoxies, standard UHU plus epoxy was chosen as an alternative. For long-term applications, further research on this topic is recommended.

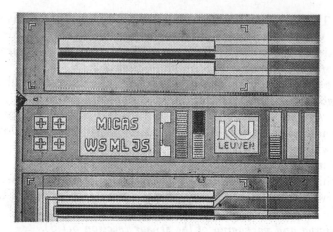

Figure 5.57 A microphotograph of a GOD–BSA membrane on a 'VOLTA I' electrode structure.

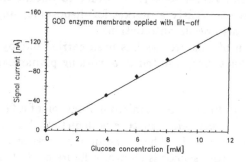

Figure 5.58 The amperometric response of a planar CMOS-compatible glucose sensor with an integrated GOD–BSA membrane.

The adhesion of the GOD–BSA membrane to the metal electrodes is not as good as the adhesion to SiO_2. This is related to the fact that there are very few hydroxyl groups available on the surface of noble metals. Two different solutions to this problem are proposed.

1. Electrochemical oxidation of the Pt electrode surface.

Pt electrodes can be oxidized electrochemically by applying a high positive voltage (1.5–2 V) to the electrode in 1M H_2SO_4. With the presently available electrode configurations, this method can only be used on packaged devices, not on a complete wafer. Therefore, a new design with a special interconnection pattern for all Pt electrodes on the wafer is required.

2. Reactive sputtering of a Pt oxide top layer.

A different solution to the adhesion problem consists of sputtering a thin Pt oxide layer on top of the Pt electrode. This can be done by changing the sputtering gas from argon to an argon–oxygen mixture during the last minutes of the Pt sputtering process. In this way, a thin layer of Pt oxides is sputtered reactively on top of the Pt electrode. The existing mask sets can be used.

It is assumed that the adhesion to Ag/AgCl electrodes is better than to Pt electrodes, since Ag/AgCl electrodes have a rough surface structure (see section 5.2). A mechanical retention of the membrane is therefore possible. However, the adhesion of the membrane to Ag/AgCl electrodes may cause problems during long-term experiments. This problem can be solved by changing the location of the membrane. In principle it is sufficient to cover only the working electrode with the enzyme membrane.

The lift-off method described above is certainly the most straightforward implementation of planar glucose oxidase membranes. It is a direct conversion of existing enzyme techniques to a CMOS-compatible microelectronic technique. Nevertheless, there are alternatives. An ideal method would be based on direct photolithography; therefore, an enzyme solution has to be prepared that can be used as a photoresist.

Some alternative solutions are the following.

1. Entrapment or cross-linking in a photosensitive hydrogel

As discussed in section 4.4.3, a photosensitive hydrogel material can be prepared by adding photosensitizers to water-soluble polymers such as polyvinylalcohol (PVA) or polyhydroxyethylmethacrylate (PHEMA). The photosensitizer used ($(NH_4)_2Cr_2O_7$) is, however, not compatible with enzymes because of the strongly oxidizing characteristic. Therefore, other types of photosensitizer are necessary.

2. Direct covalent coupling on the sensor surface.

It also possible to covalently couple the enzyme glucose oxidase to the surface of an electrode. Therefore, the electrode surface is treated in an identical way to that described in section 4.4.4. Instead of casting the BSA–GOD solution over the wafer, the sensor is placed in a GOD–PBS solution for 30 minutes. The enzyme is then coupled with the glutaraldehyde at the surface of the electrode.

This method was evaluated in Claes and Maene (1987) on thick-film sensors, but it did not give the expected results; no glucose sensitivity was observed. The amount of enzyme that can be coupled on a sensor surface is insufficient for a good glucose sensitivity and the electrode–enzyme distance is too large for direct electron transfer. However, by modification of the enzyme by the incorporation of electron relays centre (Degani and Heller 1987), this method can also result in a working glucose sensor. As the enzyme layer is very thin, lift-off will be no problem.

3. Electrochemical immobilization and conducting polymers.

An interesting method for electrochemical immobilization of enzymes is described in Bartlett and Whitaker (1987). Electrochemical polymerization is indeed an attractive technique: it allows a precise control over the amount and location of the deposited enzyme. This method is based on the immobilization of glucose oxidase in electrochemically grown films of poly-n-methylpyrrole. By controlling the current and potential, the thickness of the film can be varied easily. No patterning is necessary, as the film only grows on the electrode. This method can also be used on wafer level by interconnecting all working electrodes on the wafer. The combination of electrochemical immobilization with mediators is described in Dicks *et al* (1989).

4. Plasma etching of bovine serum albumin membranes.

Although plasma etching of enzyme membranes has not yet been reported or tested, it is believed that this method can also be applied to the patterning of planar membranes. The most important problem to be expected when etching enzyme membranes is heating of the substrate and possible destructive effects of UV radiation in the plasma. Excessive heat and UV radiation will diminish the enzyme activity.

All the methods cited above, however, have one important disadvantage; they are experimental techniques that do not guarantee a working device. The bovine serum albumin cross-linking technique, on the other hand, is well documented and almost always results in a working device. It can, however, be expected that the new methods can result in even better or more reliable devices after optimization of the process.

Alternatives also exist for the diffusion membrane. During the last few years the point of interest has moved from polyurethane membranes to Nafion membranes. Nafion is a teflon-like material with proton conducting characteristics used for hydrogen sensors. At the 'Transducers '89' conference in Montreux several authors proposed the use of Nafion instead of polyurethane (Turner *et al* 1989, Buxbaum *et al* 1989). This is mainly related to reproducibility problems involved with polyurethane membranes. Nafion solutions exist commercially and can be applied by spin or dip coating. A similar plasma etching technique to that used for polyurethane is a possible candidate as a patterning technique for Nafion membranes.

5.3.3   Oxygen sensors

An improved Clark cell is used for sensing dissolved oxygen. A major contribution of the authors is the implementation of a three-electrode configuration instead of the classical two-electrode configuration, as already discussed in section 2.3.2. The most important advantages of a three-electrode configuration in comparison with a two-electrode configuration are the elimination of electrode material consumption during the oxygen reduction and improved potential control by the elimination of polarization at the reference electrode.

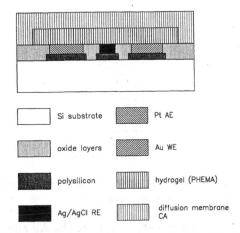

Figure 5.59 The cross section of a planar three-electrode oxygen sensor.

Although the electrode configurations discussed in section 5.1.2 are already sensitive to oxygen, in practical applications two membranes are necessary (figure 5.59). The top membrane is necessary to prevent electrode poisoning and to diminish the flow dependence of the sensor. The bottom membrane consists of a hydrogel and replaces the electrolyte of the Clark cell. The application of membranes on the existing electrode configurations has been investigated in van Hove and Placke (1988). The results of this work are summarized in the next sections.

5.3.3.1 Preparation. The preparation of planar dissolved-oxygen sensors has up to now been based on dip-coating techniques, since the study of photolithographic hydrogel materials is not yet in a definitive phase. The preparation consists of the following basic steps.

1. Processing of the CMOS-*compatible electrode structure (section 5.1).*
Not all electrode configurations can be used for oxygen sensing. Au reacts with the Cl^- ions in PBS when a positive potential is applied to the electrode, so Au electrodes are not suitable for use as auxiliary electrode. Under these conditions, there is no advantage in choosing a three-electrode configuration, as the Au of the auxiliary electrode will also be consumed during the oxygen reduction. For a three-electrode oxygen sensor, a Pt auxiliary electrode is preferred.

2. Dip coating or photolithography of a PHEMA *hydrogel (section 4.4.3).*
The PHEMA hydrogel is applied to the sensor by dip coating. The same preparation as described in section 4.4.3 is used. After dip coating, the hydrogel material is exposed to UV light to polymerize the gel and to obtain good adhesion of the gel material to the sensor surface. For good adhesion

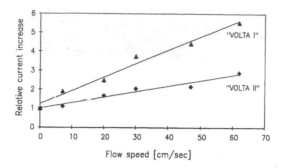

Figure 5.60   A comparison of the flow dependence of the 'VOLTA I' and the 'VOLTA II' electrode structures.

to the electrodes, it is necessary to oxidize the Au and Pt electrode surfaces electrochemically prior to membrane deposition.

3. Dip coating of a cellulose acetate oxygen diffusion membrane.

A cellulose acetate solution is prepared by dissolving 100 mg of cellulose acetate in 3 ml acetone. This solution is dip coated on the sensor, resulting in a cellulose acetate membrane with a thickness less than 5 μm.

This process results in a working dissolved-oxygen sensor. Sensitivity, selectivity, stability and reproducibility of this sensor are good.

5.3.3.2   Results.   One of the important characteristics of 'in vivo' oxygen sensors is the flow dependence of the sensor. This flow dependence has to be as low as possible. Two ways to achieve this goal are the use of microelectrodes or ultra-microelectrodes and diffusion membranes.

It is known from the literature that the flow dependence of a small electrode is less than for a larger electrode (Siu and Cobbold 1976). This is related to the thickness of the diffusion layer. As indicated in figure 5.60, the relative current increase of the 'VOLTA II' electrode structure due to flow dependence is only half the current increase of the 'VOLTA I' electrode structure. This is related to the smaller electrode width of the 'VOLTA II' working electrodes (10 μm) in comparison with the 40 μm wide 'VOLTA I' working electrodes. If the electrode width decreases, then the thickness of the diffusion layer also decreases. The disturbance of the diffusion layer and the current due to convection are therefore lower as well.

The flow dependence can also be diminished by using a diffusion membrane. If the membrane is thick enough, it can be assumed that the diffusion process is completely contained in the membrane, so the influence of the solution flowing outside the membrane (external diffusion) can be eliminated. However, as this diffusion membrane also slows down the diffusion of oxygen to the electrode, the sensitivity and the response time of the sensor will decrease. A compromise has to be found between good sensitivity, low flow dependence and fast response (see also table 2.2).

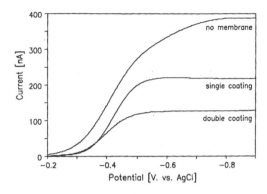

Figure 5.61 The influence of a cellulose acetate membrane on the linear-sweep voltammogram of an oxygen sensor.

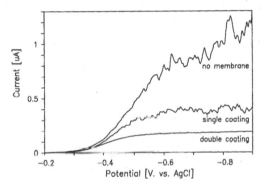

Figure 5.62 The influence of a cellulose acetate membrane on the flow dependence of an oxygen sensor (flow= 42 cm $^{-1}$s).

The influence of a cellulose acetate membrane is depicted in figures 5.61 and 5.62. A single coating on a 'VOLTA I' sensor results in a decrease in sensitivity of 41% and a decrease in the flow dependence of 60%. A double coating on a 'VOLTA I' sensor results in a decrease in sensitivity of 68% and a decrease in the flow dependence of 77%. It is also seen that the linear-sweep voltammogram of a double-coated sensor in a stirred solution approaches the shape of the linear-sweep voltammogram in a quiescent solution. These measurements indicate that a double coating of cellulose acetate provides the most interesting results. The thickness of these diffusion membranes is then of the order of 10 μm.

In this experiment, no hydrogel material is applied beneath the diffusion membrane, so a shift in E^0 value is seen on the voltammograms in a quiescent solution. The shape of the voltammogram of a coated sensor corresponds better to the theoretical shape if compared with the shape of the voltammogram of the naked sensor. It appears that the reduction process

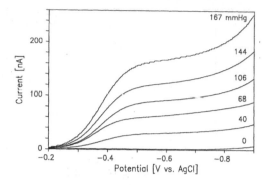

Figure 5.63   Linear-sweep voltammograms of a complete oxygen sensor for different pO_2 concentrations.

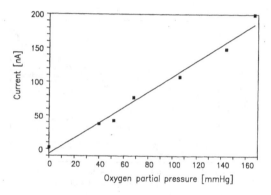

Figure 5.64   Calibration curve of a complete oxygen sensor.

is more reversible with a cellulose acetate membrane.

In figure 5.63, the linear-sweep voltammograms of a complete oxygen sensor (PHEMA hydrogel + CA membrane) are shown for different pO_2 values. The current in a nitrogen-saturated solution is almost zero. The shape of the voltammograms is regular, but no real plateau region is observed. From these voltammograms, the calibration curve can be obtained (figure 5.64). Taking into account the precision of the gas mixing equipment ($\pm 5\%$), it can be stated that the response of the sensor is linear.

In figure 5.65, the amperometric response of a complete oxygen sensor is pictured. The stability of the sensor is excellent; after some initial current increase, the current remains stable to within 4%. However, an oscillating, noisy behaviour is observed. The exact cause of these oscillations or noise is not known. The most probable cause is electromagnetic interference from the measurement equipment or $1/f$ noise.

It can be concluded that a working oxygen sensor has been made. Preliminary results of measurements in whole bovine blood (Van Hove–

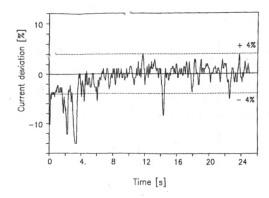

Figure 5.65 The stability of a complete oxygen sensor over 7 hours.

Placke 1988) indicate that this sensor also works in whole blood. Therefore, with some improvements of the process, a cheap disposable dissolved-oxygen sensor for 'in vivo' measurements can be realized; steps concerning further optimization are detailed in the next section.

5.3.3.3 Problems. Before this planar oxygen sensor can be used for 'in vivo' experiments on humans, some points still have to be improved. In order to realize a cheap disposable sensor, the dip-coating steps in the process have to be converted into a planar microelectronic process. This is no problem for the hydrogel layer, as planar PHEMA membranes have already been realized (see section 4.4.3). For the diffusion membrane, another material has to be found, as cellulose acetate membranes are difficult to apply with planar techniques; only screen printing seems to be an acceptable solution. It is also difficult to modify cellulose acetate into a biocompatible material because of blood clotting problems. Alternative membrane materials are commercial photoresists, teflon or PVC membranes.

The major problem remains biocompatibility. Biocompatibility does not only mean the use of non-toxic materials, but also requires a well defined interface between sensor and blood. Blood clotting and tissue growth have to be avoided under all circumstances. Incorporation of antithrombogenic material is therefore of critical importance for the future 'in vivo' use of planar oxygen sensors (Brinkman 1989).

5.3.4 CMOS interface circuits for voltammetric sensors

One of the main advantages of the generic CMOS-compatible sensor technology developed by the authors is the facility to integrate the sensor interface electronics with the sensor on the same substrate. If some signal conditioning is added to the sensor, then the device is named 'a smart sensor'. The development of an 'in vivo' smart sensor suitable for oxygen or

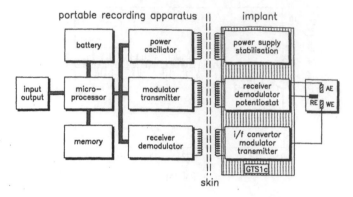

Figure 5.66 Schematic diagrams of the bi-directional telemetry system 'GLUCOSTAT'. The interface chip 'GTS1c' is the heart of the implanted system.

glucose sensing is the ultimate goal of our research on planar voltammetric sensors.

The first CMOS interface circuit with an integrated potentiostat and telemetry system has been developed by Jef Celen in our laboratory (Sansen *et al* 1983, Sansen and Lambrechts 1985). This implantable system was developed for combined use with the needle-type glucose sensor shown in figure 2.29. However, this system was not intended as an integrated interface circuit for a planar CMOS-compatible sensor. An external microprocessor unit controls the implanted circuit via a bi-directional telemetry system, as depicted in figure 5.66. The CMOS interface circuit GTS1c consists of three blocks:

1. power supply stabilization;
2. receiver, demodulator, potentiostat;
3. current–frequency converter, modulator, transmitter.

The implanted system consists of the CMOS interface circuit GTS1c mounted on a hybrid substrate, three coils for power and data transmission and the needle-type glucose sensor. The mounted hybrid is shown in figure 5.67. A similar telemetry system with reduced input communication facilities is described in McKean and Cough (1988). His system is built up with commercially available CMOS components and packaged in a pacemaker-like housing. It has two input channels, since it is intended for an oxygen detector based glucose sensor.

Although the GTS1c circuit works according to the specifications, it is too complex for integration with an 'in vivo' planar CMOS-compatible glucose sensor. Therefore, a minimum interface configuration for the realization of a smart glucose sensor has been defined (Lambrechts 1985); this configuration is shown in figure 5.68. It was developed according to the idea

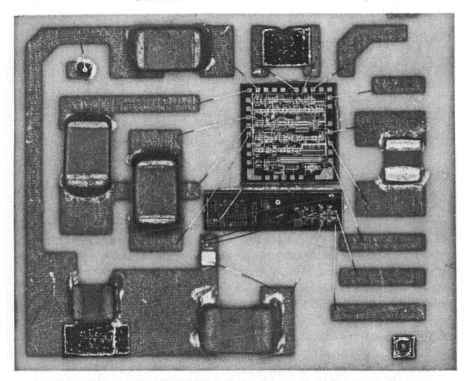

Figure 5.67 Photograph of the mounted hybrid 'GTS1c'; a glucose telemetry system.

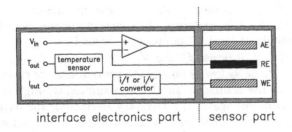

interface electronics part | sensor part

Figure 5.68 Block diagram of the most basic smart voltammetric sensor interface.

that the minimum amount of electronics needed for the conversion of the weak sensor signal to a plain low-impedance signal should be placed on the sensor itself. Complicated signal processing, such as temperature correction or calibration, is better done in an external microprocessor-controlled unit. In this way the development cost, the production cost and the area

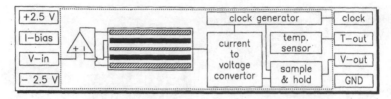

Figure 5.69   Block diagram of the smart voltammetric sensor realized for glucose and oxygen sensing.

of the circuit are minimized. There is no need yet for the integration of complex microprocessor-like circuits on the same chip as the sensor; this would decrease the yield and the flexibility of the device.

This basic configuration was further improved to the block diagram shown in figure 5.69 and realized in a p-well process by Koen Vriens (Sansen 1987). A re-design in an n-well process with improvement to the stability of the potentiostat operational amplifier and correction of the temperature sensor was done by Dirk De Wachter and Ludwig Callewaert. In what follows, this design is briefly described; for more details the reader is referred to (De Wachter 1990).

The following specifications are anticipated.

1. Current measurement: 20 nA full scale to 500 nA full scale.

Since the circuit is intended for oxygen sensors as well as glucose sensors, a broad range of currents has to be covered. This implies that an external adjustment of the full-scale value must be available. For glucose sensors, a negative current has to be measured; for oxygen sensors, a positive current has to be measured. The overall accuracy of the current measurement has to be better than 2.5%. As an output signal, a low-impedance voltage signal between −1 V and +1 V is preferred. The current measurement has to be fast enough to obtain at least 10 samples per second.

2. Power supply: −2.5 V, 0 (ground), 2.5 V.

The power supply has to be as low as possible to prevent leakage currents from the Si substrate to the solution. To obtain a sufficiently high compliance voltage (maximum voltage at the auxiliary electrode), the output of the potentiostat operational amplifier has to reach the supply voltages as closely as possible. The maximum power consumption of the interface chip has to be lower than 2 mW. The power supply has to be galvanically isolated from the patient externally to prevent electrocution in case of a circuit malfunction.

3. Potentiostat operational amplifier.

The potentiostat circuit has to remain stable for all possible sensor configurations (glucose or oxygen sensor). The phase margin with a 10 nF load has to be larger than 60°. The gain–bandwidth product of the operational amplifier has to be around 100 kHz. The offset has to be smaller than

10mV. The output of the potentiostat operational amplifier has to reach the supply voltages as closely as possible. The input potential can vary between −1 V and +1 V. To prevent electrical stimulation of the patient and tissue damage when the reference electrode fails, the maximum output current has to be lower than $5\mu A$.

4. Temperature sensor.

The sensor must operate between 15 °C and 42 °C. To be allowed to display the temperature on a monitor, an accuracy better than 0.05 °C has to be obtained.

5. Dimensions.

To fit into a French 5 catheter (external diameter = 1.65 mm), the width of the chip has to be smaller than 0.7 mm. The length of the chip, including the sensor may not exceed 6 mm. The total number of bonding pads is preferably four.

6. Electromagnetic interference.

For 'in vivo' use, the circuit has to resist internal defibrillation shocks (40 J) and high-frequency surgery.

The resulting block diagram consists of the following (figure 5.68).

1. Potentiostat operational amplifier.

The potentiostat operational amplifier has been designed for optimum stability with a 10 nF load (Britz 1978). To obtain a wide output swing, a pMOS source-follower output configuration has been chosen.

2. Current-to-voltage convertor.

The current-to-voltage convertor is based on a switched capacitor integrator with an on-chip capacitance of 10 pF. By changing the clock frequency the current range can be adjusted between 20 nA (1 kHz) and 500 nA (25 kHz) full scale.

3. Sample-and-hold circuit.

The obtain a DC output signal, the current-to-voltage convertor is followed by a sample-and-hold circuit.

4. Clock generator.

The clock generator creates all necessary internal clock signals for the current-to-voltage convertor and for the sample-and-hold circuit from the external clock signal.

5. Temperature sensor.

The temperature sensor is based on the temperature dependence of a CMOS voltage reference, where two pMOS transistors are biased in weak inversion (Hosticka *et al* 1984). The output voltage is a linear function of the absolute temperature (PTAT).

6. Voltammetric sensor.

The voltammetric sensor configuration is identical to the VOLTA II structure (see section 5.1.2). A two-electrode as well as a three-electrode configuration have been designed. The interface circuit is made compatible with

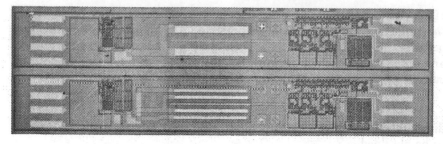

Figure 5.70 Microphotograph of the smart sensor interface chip, shown after sensor processing.

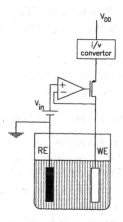

Figure 5.71 Schematic diagram of a simple integrated potentiostat described in Turner R F B *et al* (1987).

both types of electrode configuration.

This design was evaluated and functioned according to the specifications (De Wachter 1990). In figure 5.70, a microphotograph of the sensor interface chip is shown after sensor processing. A two-electrode oxygen sensor (WE=Au, RE=Ag), as well as a three-electrode glucose sensor (WE=AE=Pt, RE=Ag), have been produced.

This smart voltammetric sensor is the first integrated potentiostat circuit with on-chip electrochemical electrodes and an on-chip temperature sensor in the world. Because of the extremely small width of the sensor, the device is suitable for catheter-tip packaging and, therefore, for 'in vivo' use. Also, the interface circuit is versatile, so that it can be used for glucose sensors as well as oxygen sensors. With the range-selectable current-to-voltage convertor, small currents (30 nA full scale) and large currents (500 nA full scale) can be measured.

A CMOS potentiostat for amperometric sensors is also described in Turner R F B *et al* (1987). This device is used as an external interface

Table 5.5 A comparison between the integrated potentiostat described in Turner (1987b) and our smart sensor design.

characteristics	[Turner-87]	our design
smart sensor	no	yes
integrated electrodes	no	yes
on-chip temperature sensor	no	yes
electrode system	two	three or two
differential configuration	yes	no
process	5 μm CMOS	3μm CMOS n-well
chip dimensions	NA	0.75 mm x 4.8 mm
area electronics	0.53 mm²	1.5 mm²
supply voltage	+/- 5 V	+/- 2.5 V
power consumption	< 2 mW	1.7 mW
potentiostat opamp	two-stage	two-stage
open-loop gain	90 dB	100 dB
unity-gain bandwidth	4 MHz	6 kHz
phase margin	NA	> 70° (10 nF load)
bias current	3.5 mA	1 μA
WE bias voltage	internal	external
bias voltage offset	< 10 mV	< 10 mV
i/v convertor	resistor	SC-integrator
range-selectable	no	yes
sensitivity	60 mV/μA	33 mV/nA (30 nA FS)
current range	0.01-3.5 μA	30 nA FS - 1 μA FS
nonlinearity	2 %	0.3 %
reproducibility	NA	0.25 % of reading
overall accuracy	NA	2%

NA : not available   FS : full scale

circuit for a Pt wire glucose sensor. It is a two-channel circuit for two-electrode systems. A schematic diagram is shown in figure 5.71. The current-to-voltage conversion is accomplished simply by mirroring the electrode current through an on-chip resistance. The sensitivity of this CMOS potentiostat is only moderate (60 mV μA^{-1}), if compared with the sensitivity of our design in the most sensitive range (50 mV nA^{-1}). In his text Turner even reports a sensitivity of 60 mV mA^{-1}, but it must be assumed that this is a typing error, when compared with his experimental results. The applied potential is fixed and can only be changed by modulating the supply voltage. Some more characteristics are summarized and compared with our circuit in table 5.5.

Turner also does not see any advantage in the integration of the sensor with his potentiostat on the same substrate. He states: '*Full integration of the transducing electrodes (including the reference) with the controlling circuitry cannot be accomplished with a standard* CMOS *process since the usual aluminium metallization layer does not provide an electrochemically active electrode surface. Some additional processing is therefore required in order to make use of this or any integrated circuit as an amperometric sensor. However, the inherent noise immunity of the amperometric technique allows this circuit to be used to control externally connected electrodes*'.

It is clear that Turner's idea about integration of sensor interfaces is completely opposite to the ideas presented in this book. Our main objective was to develop a CMOS-compatible sensor process so that integration of the interface electronics with the sensor is self-evident. It should also be noted that amperometric sensors are only immune to noise when large-area sensors are used. For 'in vivo' applications, small sensors with a small electrode area are necessary; this small electrode area is also necessary to reduce the flow dependence. It is not possible to measure these small currents with externally connected electrodes in an electromagnetically noisy environment such as, for example, an intensive care unit or an operating room.

It can be concluded that our concept of smart voltammetric sensors is unique in the world. It combines a well studied electrode configuration with state-of-the-art interface electronics. In comparison with other integrated potentiostats, superior characteristics are obtained. It is believed that with this concept an unrivalled device has been made that can be used on a commercial base for short-term 'in vivo' applications within four to eight years. Therefore, the development of this device has to be continued. Attention has now to be directed towards new packaging techniques, reduction of the number of bonding pads, elimination of electromagnetic interference and calibration procedures.

5.3.5 ESCAPE: a planar conductometric urea sensor

In order to prove that the techniques developed for glucose and oxygen sensors can be expanded to other biosensors, the development of a planar conductometric urea sensor is discussed.

For the 'in vivo' efficiency measurement of an artificial kidney treatment, a conductometric urea sensor is now in development by Paul Jacobs. This research is carried out in cooperation with Dr P Hombroux of the Kliniek Hogerlucht in Ronse (Belgium). This sensor is based on the change in conductivity beneath a urease membrane when the device is submitted to a varying urea concentration (Watson 1987).

In the presence of the enzyme urease, urea is hydrolized. The following reaction takes place:

$$NH_2CONH_2 + 2H_2O + H^{+} \overset{(urease)}{\longrightarrow} 2NH_4^{+} + HCO_3^{-}. \tag{5.2}$$

Figure 5.72   Microphotograph of the realized 'ESCAPE' structure.

Because of the generation of conducting ionic species, the conductivity of the solution increases as the urea concentration increases. Monitoring of the conductance is preferred above other detection techniques (such as an ammonium-selective ISE, a pH electrode, an ammonia gas sensor or a pCO_2 sensor) because of the fact that a conductometric cell can be fabricated easily with the microelectronic techniques described in this book.

Therefore, a planar conductometric cell, 'ESCAPE' (Electrode System for Conductometric And Potentiometric Experiments), has been designed by the authors in cooperation with Guido Huyberechts and Paul Jacobs. The ESCAPE cell is designed in such a way that the same device can also be used for gas sensor applications. As an optimum structure an interdigitated structure is chosen with fingers which are 20 μm wide and 20 μm apart. Each electrode has 20 fingers. This results in a width-to-length ratio equal to 3200. The dimension of one conductometric cell is 1.6 mm by 3.6 mm. A microphotograph of the 'ESCAPE' structure is shown in figure 5.72. Devices with Au electrodes and with Pt electrodes have been realized.

As shown in the layout of the sensor, two conductometric cells are placed on one die; this allows a differential measurement to be made. On the right-hand electrode, the enzymatic membrane is applied. This membrane consists of urease immobilized in a bovine serum albumin matrix according to the same principles as described in section 4.4.1. On the left-hand electrode, only the bovine serum albumin matrix is applied. To compensate for conductivity changes in the solution not related to urease enzyme activity, the conductivity of both cells is measured and the differential signal is determined. This differential signal is proportional to the urea concentration.

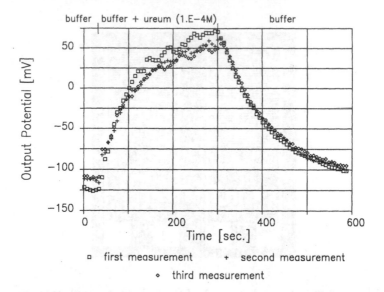

Figure 5.73   Response of the conductometric urea sensor (based on the 'ESCAPE' structure) in buffer solution.

The sensor has been evaluated by Paul Jacobs (Jacobs 1989). A preliminary response is depicted in figure 5.73. This experiment shows that conductivity changes related to enzyme activity can be measured as a function of time with the ESCAPE electrode structure. The relatively long response time is related to the suboptimal design of the flow cell. The conductometric urea sensor itself responds very quickly. The response can be improved drastically by platinization of the electrodes (Jacobs *et al* 1990).

From this preliminary experiment, it can be concluded that after some optimization of the device it is possible to realize a planar urea sensor. Because of the low production cost, this device can be used as a disposable sensor for the 'in vivo' monitoring of the dialysis process with artificial kidney patients. This will improve the quality of life of these patients substantially. At this moment, the result of the dialysis is only known a few days after the treatment. Continuous monitoring gives direct information on the dialysis process, resulting in a short and efficient dialysis.

These preliminary results also demonstrate that the generic CMOS-compatible process developed for glucose and oxygen sensors can be extended to other measurement techniques. This clearly shows that the generic process is a valuable tool for the development of planar CMOS-compatible biosensors.

A good indication of the commercial value of this research is the price of a chemiresistor microelectrode array sold by Microsensors Systems Inc (Microsensors 1988). This Microsensors Systems interdigitated electrode

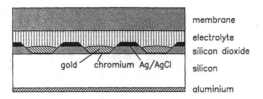

Figure 5.74 Microelectrochemical oxygen sensor based on the multi-cathode principle (adapted from Siu and Cobbold (1976)).

system is very similar to the 'ESCAPE' configuration detailed in this section. The chemiresistor measures 12.5 mm by 7.5 mm and is fabricated on quartz instead of silicon. The interdigitated pattern is made in gold; 50 fingers (20 μm wide with 20 μm spacing) are placed on one sensor. The bare electrode structure, without chemically sensitive layers, sells for $45 each (5 to 9 pieces) or $37.5 each (100+ pieces). This is an extremely high price for such a mass-produced device. The fabrication cost of one 'ESCAPE' sensor, (also without a chemically active membrane) can be rated at $0.5, taking into account a production batch of 10 wafers with 600 devices on one wafer. This corresponds to a total batch cost of $3000, including wafers, thermal oxidation, photolithography, Au evaporation, lift-off and wafer dicing.

5.4 EXAMPLES OF OTHER RESEARCH GROUPS

Dissolved-oxygen sensors based on the Clark cell principle were the first sensors realized with microelectronic fabrication technology. Microelectrochemical sensors fabricated with integrated-circuit production techniques have already been reported in the literature in 1974 by Butler and Cobbold (1974). They fabricated a multicathode oxygen sensor as a variant of the Clark cell. A theoretical treatise of this multicathode oxygen electrode can be found in Siu and Cobbold (1976). Circular Au cathodes with a diameter of 7 μm are embedded in a large Ag/AgCl anode. As shown in figure 5.74, all cathodes are interconnected via the silicon substrate. Contact of the multicathode is made with an aluminium metallization.

A miniaturized electrode for on-line pO_2 measurements is presented in 1975 by Eden *et al* (1975). This sensor measures 2 mm by 2 mm and is fabricated with microelectronic production techniques. The electrodes are recessed in a deep cavity etched in the silicon providing a reservoir for a KCl solution. This cavity has to be covered with an oxygen permeable membrane. In Eden *et al* (1975) the possibility of integrating interface electronics with the sensor was already suggested: *'The silicon substrate used for the electrode can be used also to produce the input stage of the 'electrode amplifier' simultaneously with the production of the electrode itself, thus decreasing electronic noise received in the transmission lines.'*

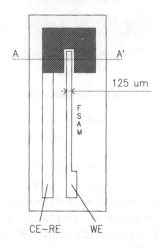

Figure 5.75 Layout of a two-electrode oxygen sensor (adapted from Koudelka and Grisel (1985)).

In the Philips Medical Systems Division in Eindhoven, The Netherlands, the first ideas of making a disposable oxygen sensor for biomedical applications using planar microelectronic techniques originated in 1976 when Kuypers reported on the 'in vivo' measurement of blood gases (Kuypers 1990). More details on this sensor and its further evolution are discussed in section 5.4.1.

A glucose sensor using anisotropic etching techniques is described in Miyahara *et al* (1983). It is a miniaturized Clark cell covered with a GOD enzyme membrane. This device is sometimes incorrectly referred to as the first microamperometric or microelectrochemical sensor fabricated with microelectronic production techniques.

At the 'Transducer '85' conference, the authors reported their planar techniques for the 'GLUCO' sensor (Sansen *et al* 1985). At the same conference, Koudelka presented a two-electrode planar 'Clark-type' oxygen sensor (Koudelka and Grisel 1985, Koudelka 1986). On top of an Ag cathode and an Ag/AgCl anode a PHEMA hydrogel layer and a silicone rubber membrane are applied manually. After packaging the sensor, these membranes are respectively cast and dip coated. The layout and cross section of this device are shown in figures 5.75 and 5.76. A further evolution of the work of Koudelka at IMT, Neuchatel is sketched in section 5.4.2.

At the Bordeaux conference on chemical sensors the authors presented their work on corrosion resistance of Ti/Pd/Ag/AgCl layers. Several Japanese groups described planar microelectrochemical sensors fabricated with microelectronic fabrication technology. A special miniaturized chamber-type electrochemical cell for medical applications is described in Prohaska *et al* (1986).

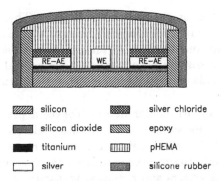

silicon	silver chloride
silicon dioxide	epoxy
titanium	pHEMA
silver	silicone rubber

Figure 5.76 Cross section of the two-electrode oxygen sensor shown in figure 5.75 (adapted from Koudelka and Grisel (1985)).

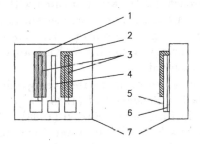

Figure 5.77 A differential two-electrode glucose sensor (adapted from Moriizumi *et al* (1986)): 1. PVA with active GOD; 2. PVA with deactivated GOD; 3. anode; 4. common cathode; 5. Au (150 nm); 6. Cr (10–30 nm); 7. glass plate (10mm × 10mm).

A differential two-electrode glucose sensor is demonstrated in Moriizumi *et al* (1986). The configuration of this sensor is shown in figure 5.77. Three gold electrodes are used: two Au working electrodes and one Au combined reference/auxiliary electrode. One working electrode is covered with a PVA–GOD mixture. The other working electrode is covered with an identical PVA layer but with deactivated GOD enzyme; this electrode provides a reference signal. A photosensitive PVA layer was used which was based on the use of a photosensitizer with stilbasolium groups. Using a differential set-up, interference from ascorbic acid and uric acid can be eliminated.

A very simple two-electrode glucose sensor is presented in Karube and Tamiya (1986). Two identical Au microelectrodes are used: one as a working electrode, the other as a combined reference/auxiliary electrode. The same sensor is also used as an oxygen detector in an L-glutamate sensor.

It is important to notice that all these authors still use a two-electrode measurement set-up, limiting the lifetime of the sensor. The working electrodes for H_2O_2 detection are even made in gold. This implies that when making measurements in a solution containing Cl^- ions, the working electrodes will chloridate and dissolve. As all biological fluids contain Cl^--ions, these sensors are not compatible enough in a biomedical environment. The sensor processing is also not adapted for further integration with CMOS or bipolar technology.

At the 'Transducer '87' conference in Tokyo, a planar three-electrode glucose sensor with Pt working and auxiliary electrodes was described (Koudelka *et al* 1987). More details on this work can be found in section 5.4.2. A multiple chamber-type probe consisting of several temperature sensors, oxygen sensors and potential transducers is explained further in Prohaska *et al* (1987). A combined ISFET urea sensor and amperometric glucose sensor is introduced in Murikami *et al* (1987) and Kimura *et al* (1988). In this approach, ISFETs are fabricated with silicon-on-sapphire (SOS) technology. An ISFET is used as a urea sensor and as reference electrode for the amperometric glucose sensor. The enzyme membranes are applied with lift-off technology, according to Kuriyama *et al* (1986). Au is still used as electrode material.

An oxygen and glucose sensor fabricated using micromachining techniques in Si is described in Suzuki *et al* (1988): a patent application for this sensor is decribed in Karube and Suzuki (1988). The sensor is rather large: the width measures 2, 3, or 4 mm and the length equals 15 mm. A two-electrode system with two Au electrodes is used. An agarose gel containing 0.1 M KCl is poured into the etched cavities of the sensor; the thickness of this layer equals 300 μm. A negative photoresist (2 μm thick) is spin coated to form the oxygen diffusion membrane.

An interesting alternative for the deposition of enzyme membranes is proposed in Tamiya *et al* (1989). In this paper a planar glucose sensor using electron mediators immobilized on a polypyrrole-modified electrode is described. Gold electrodes were coated with polypyrrole by electropolymerization. Glucose oxidase is adsorbed on this polypyrrole film by dipping in an aqueous GOD solution overnight. As a mediator, dimethylferrocene is used. This mediator is dip coated in a polyvinylbutyral matrix on the GOD-activated sensor. A wide dynamic range of the glucose sensor in combination (0–30 mM) with a low applied potential are achieved.

At the 'Transducer '89' conference in Montreux, further developments of the previously described designs were presented. The authors presented the smart sensor for the voltammetric measurement of oxygen or glucose concentrations described in section 5.3.4. A three-electrode system consisting of circular concentric electrodes is introduced in Pilf *et al* (1989). A circular Ag/AgCl electrode is surrounded by circular Au working and auxiliary electrodes. Although the sensor is intended for 'in vivo' use, interference of Cl^- ions in the solution with the Au electrodes has not been taken into account. A planar glucose sensor was presented by the same group using Nafion as a selective diffusion-restricted membrane. This sensor is fabricated on a glass substrate.

A further evolution of the chamber-type sensor is sketched in Schneider *et al* (1989). A new design of the micromachined device from Suzuki *et al* (1988) is presented in Suzuki *et al* (1989). With an identical structure, several biosensors for the detection of glucose, carbon dioxide and L-lysine have been realized. All these biosensor are based on the measurement of dissolved oxygen in the enzymatic or microbial membrane. Further details are given in Suzuki *et al* (1990, 1991). In Morff *et al* (1990) it is shown that non-semiconductor substrates, namely Kapton, can also result in a glucose sensor with potential 'in vivo' applications. In Gumbrecht *et al* (1991), the 'ex vivo' operation of a three-electrode oxygen sensor mounted at the back of a multi-lumen catheter is demonstrated. A PHEMA membrane on Pt electrodes is used in combination with a special regeneration potential waveform.

This historical overview of planar microelectrochemical sensors can be concluded by the remark that the interest in this type of sensor is increasing steadily. In comparison with ISFETs, however, only a limited amount of research has been, or is being, done. Practically all the work on microelectrochemical sensors deals with voltammetric or amperometric devices. A planar conductometric sensor is described in Watson *et al* (1987) and Sheppard (1991). Planar potentiometric sensors are also not so intensively studied. The most important papers dealing with this topic originate from the I-STAT company, formerly Integrated Ionics (see section 5.3.1).

Some microelectrochemical sensors that are closely related to the work of the authors are described in the next sections.

5.4.1 A disposable catheter-tip oxygen sensor

One of the microelectrochemical sensors that is very close to a commercial introduction is the dissolved-oxygen sensor developed by Kuypers. This development started in 1976 at the Philips Medical Systems Division. The first device was made in 1978 in the Honig and Kuypers (Kuypers 1990, Honig 1990). A schematic diagram of this device with a working electrode in the form of a spiral is shown in figure 5.78 (type 1) (see also Honig 1980).

A small research team continued this interesting work. This research team is headed by Martin Kuypers of Dräger Medical Electronics (Best, The Netherlands) who has been working in the field of microelectrochemical pO_2 sensors since 1976. Although this company has changed owner and name several times (Honeywell and Philips, Honeywell, PPG Hellige and now Dräger Medical Electronics), the development of a catheter-tip 'in vivo' dissolved-oxygen sensor remained an important research activity.

This sensor, a miniature two-electrode Clark cell, is manufactured on a production line for discrete bipolar electronic components. The processing

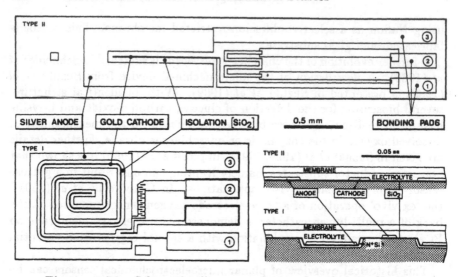

Figure 5.78   Schematic diagrams of type I and type II dissolved-oxygen sensors developed by the Honeywell and Philips Medical Electronics Group in Best, The Netherlands (adapted from Engels and Kuypers (1983), with courtesy of M H Kuypers, Dräger Medical Electronics (Best, The Netherlands).

is therefore a nice example of compatibility with bipolar IC production techniques.

A schematic diagram of the pO_2 sensor, as described in Engels *et al* (1982) and Engels and Kuypers (1983), is shown in figure 5.78. In the type I device, efforts were made to embed the anode in the silicon, providing an electrolyte reservoir. The anode and cathode were made in the form of a spiral. The major disadvantage of this design is the small area of the anode compared to the cathode, restricting lifetime of the sensor to approximately 4 hours. In the second design these drawbacks have been corrected, resulting in a sensor with an operational lifetime of 2 days. This lifetime restriction is due to the limited amount of Ag present on the anode. This problem can be dealt with by the introduction of a third electrode: whereas the authors prefer an integrated Pt auxiliary electrode (see section 2.3.2), an external auxiliary electrode is proposed in Kuypers (1988). A temperature-sensing element was incorporated in both designs and is used to compensate the temperature sensitivity of pO_2 sensors. The electrolyte membrane and oxygen diffusion membrane were applied by spin coating and photolithography. The electrolyte layer consists of a photosensitive hydrogel material, namely polyvinyl pyrrolidone. The diffusion membrane is realized with a negative photoresist consisting of cyclo-cis-isoprene. The electrolyte layer is rehydrated very quickly by providing small holes in the diffusion membrane (Kuypers 1985). Placement and dimensions of the

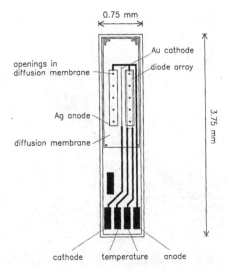

Figure 5.79 Layout of the new design of the dissolved-oxygen sensor (courtesy of M H Kuypers, Dräger Medical Electronics (Best, The Netherlands).

holes are chosen in such a way that diffusion of oxygen through the diffusion membrane is predominant compared to diffusion of oxygen through the small holes. This sensor has, for example, been used with success for the on-line measurement of conjunctival oxygen tension (van der Zee *et al* 1985).

A further improvement of this design is shown in figure 5.79. For temperature sensing, diodes are placed under the active part of the sensor. The dimensions of this sensor are adapted for catheter-tip packaging: the working electrode is only 5 μm wide and approximately 1 mm long. In figure 5.80, a pO_2 sensor packaged in a stainless steel catheter tip is shown.

Special attention has been paid by this research group to biocompatibility of their devices. Biocompatibility is increased, for example, by incorporating heparin into a PVA top layer (Kuypers *et al* 1988, Brinkman 1989). This PVA membrane has the following characteristics. It was cross-linked in such a way that a semi-permeable membrane was formed, preventing the diffusion of large molecules in the sensor. In this way poisoning of the electrodes and interference of other reducible molecules is avoided. Most of the heparin is covalently bound tho the PVA matrix; a small amount of heparin, however, is allowed to migrate out of the membrane resulting in a micro-antithrombogeninic region close to the membrane. Therefore, successful 'in vivo' experiments on dogs were performed. By combining a well designed catheter-tip package with a biocompatible sensor top layer, a pre-commercial device has been realized.

Recent interest in glucose sensing has also resulted in a collaboration

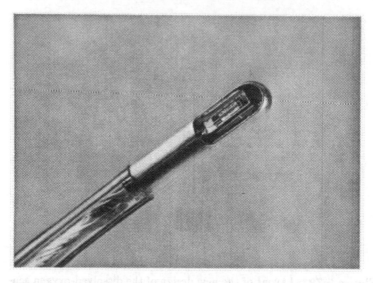

Figure 5.80   Catheter-tip package of a dissolved-oxygen sensor (courtesy of M H Kuypers, Dräger Medical Electronics (Best, The Netherlands).

between the authors and the Dräger Medical Electronics group on glucose sensing techniques under anaerobic conditions (Sansen *et al* 1991, Kuypers and Steegs 1989). A feasibility study completed at K U Leuven in the ESAT–MICAS group showed that, by the introduction of a potential pulse at the working electrode, enough oxygen is generated at this working electrode to allow the functioning of a GOD enzyme based sensor under anaerobic conditions. This principle is demonstrated by the following amperometric experiment. The potential profile sketched in figure 5.81 is applied to a conventional H_2O_2-based glucose sensor. No mediators are present in the enzymatic membrane.

Two different operational phases can be distinguished in figure 5.81. In the first phase, the working electrode is pulsed to a high potential (1–1.2 V) where dissociation of the water molecules occurs. During this dissociation, oxygen is generated. This oxygen is then consumed during the enzymatic reaction, resulting in the generation of H_2O_2. This H_2O_2 is subsequently measured amperometrically in the second phase by lowering the potential. In figure 5.82, the results of this novel measuring technique are shown. In a PBS solution saturated with nitrogen, no response to an increasing glucose concentration is seen. By applying a potential pulse, however, a linear response to a varying glucose concentration is seen.

A similar experiment, but without a GOD enzyme membrane on the sensor, did not show the same effect. No glucose-related current was observed. This proves that the pulse effect is related to oxygen generation and not

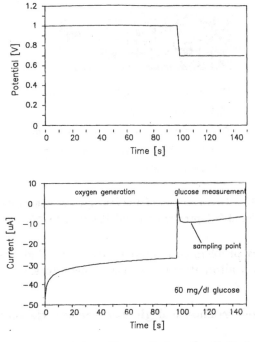

Figure 5.81 Potential profile applied to the H_2O_2-based glucose sensor for operation under anaerobic conditions and the measured current.

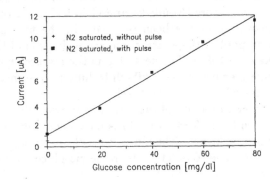

Figure 5.82 Response of a conventional glucose sensor in PBS saturated with nitrogen (anaerobic conditions). Influence of in situ oxygen generation by intermittent pulsing of the working electrode to high potentials.

to activation of the electrode surface by pulsing techniques, such as those described in section 2.3.2 (pulsed amperometric detection).

It is believed that this new method is an interesting solution to the

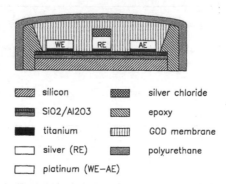

silicon		silver chloride	
SiO2/Al2O3		epoxy	
titanium		GOD membrane	
silver (RE)		polyurethane	
platinum (WE–AE)			

Figure 5.83   Cross section of a planar glucose sensor developed at IMT, Neuchâtel (adapted from Koudelka *et al* (1987)).

oxygen deficiency problem encountered with 'in vivo' glucose sensors. It is certainly a worthy alternative to the use of mediators.

5.4.2   Microelectrochemical glucose sensors developed at IMT, Neuchâtel

The development of a microelectrochemical glucose sensor at the Institute of Microtechnology in Neuchâtel, Switzerland, followed a similar method to the research performed by the authors. Based on experience on microelectrochemical pO_2 sensors (Koudelka and Grisel 1985), a three-electrode H_2O_2-based glucose sensor has been realized. In the first phase, special attention has been paid to the electrochemical evaluation of thin-film electrodes (Koudelka *et al* 1987). The next step in the research was the deposition of the enzyme membrane using lift-off techniques (Gernet *et al* 1989a) similar to those described in section 4.4.4.

In figure 5.83, a cross section of the device is shown. The corresponding layout is depicted in figure 5.84. No details have been published about the motive for this unusual layout. The dimensions of the working electrode are 100 μm by 1000 μm. The complete amperometric glucose sensor measures 0.8 mm by 3 mm.

The sensor process can be summarized as follows (Gernet *et al* 1989b).

1. *Thermal oxidation of the silicon wafer (150 nm)*
2. *Deposition of an Al_2O_3 layer (150 nm)*

This layer is deposited by CVD techniques in order to increase the lifetime of the device. The thin thermal oxide has poor chemical stability. Al_2O_3 layers are known to be excellent passivation layers. Compared to Si_3N_4, Al_2O_3 has the advantage that membrane adhesion is improved.

3. *Pattern formation of the Pt layer*

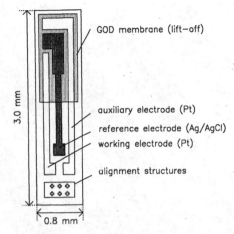

Figure 5.84   Layout of a planar glucose sensor developed at IMT, Neuchâtel (adapted from Koudelka *et al* (1987)).

Lift-off techniques are used for the formation of the Pt working electrodes. A multilayer consisting of 50 nm Ti and 150 nm Pt is evaporated in an electron-gun evaporator.

4. *Pattern formation of the Ag layer*

The Ag layer is also formed using lift-off techniques. As an adhesion layer, 50 nm of Cr is used. The thickness of the Ag layer is 1 μm. This Ag layer is partially chloridated in a 0.025 M $FeCl_3$ solution. Corrosion resistance is obtained by proper choice of evaporation parameters (Koudelka and Grisel 1985).

5. *Deposition of the GOD membrane*

The glucose enzyme membrane, consisting of GOD, BSA and glutaraldehyde, is cast manually on the sensor. A typical composition is 50 mg ml^{-1} GOD and 80 mg ml^{-1} BSA. To this mixture, 2.5% (v/v) glutaraldehyde is added. On the packaged sensor, 10 ml of this solution is cast, resulting in a 50 μm thick GOD membrane.

For the deposition of the enzyme membrane on the entire wafer, lift-off techniques have been adopted (Gernet *et al* 1989a).

6. *Deposition of the polyurethane membrane*

The glucose diffusion membrane is formed by dip coating the sensor in 4–6% polyurethane dissolved in a 1/9 (v/v) mixture of dimethylformamide and tetrahydrofuran. The diffusion membrane is only 5 μm thick.

This process results in the sensor shown in figure 5.85.

In contrast with the work of the authors, no research has been performed on the integration of interface electronics with the sensor. This process is not directly compatible with CMOS or bipolar techniques due to the Al_2O_3 layer and the non-standard connection of the noble-metal layers. With

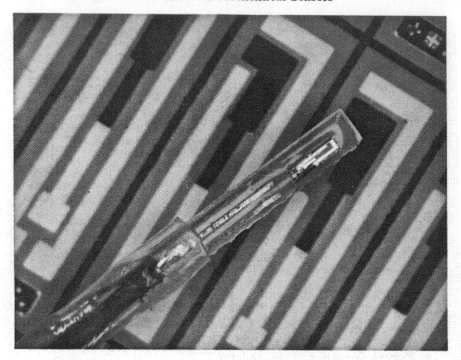

Figure 5.85 Microphotograph of the packaged glucose sensor with a processed wafer in the background. The sensor is mounted on a miniature printed circuit board (courtesy of Professor N de Rooij, IMT, Neuchâtel).

minor modifications, however, this sensor process can be converted to a process suitable for integration with interface electronics.

The research group at IMT, Neuchâtel, however, has more experience with 'in vivo' testing of their glucose sensors. Successful 'in vivo' tests have been performed in rats (Koudelka *et al* 1988, 1991). The sensor described above has also been converted to a free-chlorine sensor by photolithographically applying a pHEMA diffusion membrane (van den Berg *et al* 1991).

5.5   CONCLUSION

In this chapter, several microelectrochemical sensors are described. A generic CMOS-compatible sensor process developed by the authors at the K U Leuven is detailed. Using this sensor process, a complete series of novel sensors has been realized. A smart sensor interface for an eight-channel ion-selective sensor array has been presented. Since the main topic of this book is voltammetric sensors, special attention has been paid to the realization of planar glucose and oxygen sensors. These sensors have been combined with on-chip interface electronics and a temperature sensor. Therefore, the first smart voltammetric sensor has been realized. In order to prove the generic nature of the sensor process which has been developed, a conductometric urea sensor has also been introduced.

Based on information found in the proceedings of the last sensor conferences and journals, an historical overview of research on microelectrochemical biosensors has been compiled. Some sensor work which is comparable to the work of the authors is discussed in more detail.

Although the first electrochemical sensor prepared with microelectronic production techniques was presented in 1974 (Butler and Cobbold 1974), no commercial devices have yet reached the market. The miniature dissolved-oxygen sensor developed by the research group of M H Kuypers is perhaps, or has been, the closest to commercial introduction onto the market. The interesting point to this development is the incorporation of special biocompatible layers and well developed catheter-tip packaging. In the field of potentiometric sensors, a commercial introduction of EGFET devices developed by I-STAT is expected (see section 5.3.1).

CHAPTER 6

THICK-FILM VOLTAMMETRIC SENSORS

Thick-film technology allows a broad spectrum of materials to be screen printed with the same equipment. This implies that thick-film technology is more cost effective than integrated-circuit-compatible sensor technology for the mass production of relatively low production volumes. The investment for a thick-film facility is moderate and the operational costs are reasonable. Thick-film technology, however, has the disadvantage that extremely small dimensions, necessary for implantable sensors, cannot be realized. The minimal achievable line-width is 100 μm, so thick-film sensors are at least one order of magnitude larger than IC technology based sensors.

With thick-film technology, mainly voltammetric sensors have been fabricated (Pace 1985, Suls 1986, Karagounis 1986, Lewandowski 1987, Scholze 1991). However, potentiometric devices are also possible, such as the pH electrodes described in Pace (1986) and Belford et al (1987). In Weetall and Hotaling (1987), an extremely low-cost, immuno-assay sensor is presented which is silk-screen printed on cardboard.

A new range of planar voltammetric sensors has been developed by the authors using screen-printing technology. A completely novel electrode material in the world of planar sensors has been introduced, namely ruthenium dioxide. This electrode material has some interesting electrochemical characteristics and is ideal for the development of H_2O_2-based biosensors such as glucose sensors. Due to the limited miniaturization, thick-film sensors are not suitable for 'in vivo' applications. They are, however, the ideal planar sensors for 'in vitro' applications.

Our research on thick-film sensors has hitherto resulted in four publications: (Suls et al 1986, Paszczynski et al 1986, Lambrechts et al 1987, 1988). The world-wide interest in these publications indicates that thick-film technology applied to the development of planar voltammetric sensors is a worthy alternative to high-cost IC technology and labour-intensive manual production methods.

246

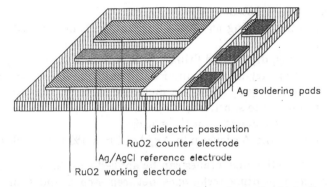

Ag soldering pads

dielectric passivation
RuO2 counter electrode
Ag/AgCl reference electrode
RuO2 working electrode

Figure 6.1 A schematic view of a three-electrode thick-film sensor. The RuO$_2$ electrode material can be interchanged with Au, Pt or any other electrode material.

Other research groups have also been working on the development of thick-film electrochemical sensors; some examples are given in section 6.3. The first commercial planar glucose sensor is also a screen-printed sensor, namely the ExacTech glucose test strip manufactured by MediSense (see section 6.3.3).

6.1 PROCESSING OF THICK-FILM VOLTAMMETRIC SENSORS

Only three basic steps are necessary for the fabrication of a thick-film layer. Firstly, the paste is applied onto the ceramic substrate with a screen printer. The printed layer is then dried and fired in a conveyor furnace. These basic steps are detailed in section 4.5. The different layers of a screen-printed sensor are all applied on the substrate using the same technique and equipment.

In figure 6.1, a schematic view of a three-electrode thick-film sensor is shown. Four different layers are necessary for the fabrication of a voltammetric electrode system: the Ag/AgCl layer for the reference electrode, the RuO$_2$ layer for the working electrode and the auxiliary electrode and a Pd/Ag layer for the soldering pads; these layers are covered and insulated with a dielectric passivation layer, with the exception of the active electrode areas and the soldering pads. Of course, the RuO$_2$ layer can be interchanged with other noble-metal layers such as Pt or Au, depending on the application.

6.1.1 The different processing steps

Based on figure 6.1, the following process sequence can be derived:

1. cleaning of the substrate (96% Al_2O_3);
2. screen-printing, drying and firing the RuO_2, Pt or Au paste for the WE and AE;
3. screen-printing, drying and firing a pure Ag paste for the RE;
4. screen-printing, drying and firing Pd/Ag paste for the soldering pads;
5. screen-printing, drying and firing dielectric paste for the passivation;
6. laser scribing of the substrate;
7. soldering of the connector on the sensor;
8. chloridation of the Ag layer into Ag/AgCl in a 1% $FeCl_3$ solution.

The chemically active membranes can be applied onto the complete substrate using screen-printing techniques between step 5 and 6 or on the individual sensors using dip-coating techniques after step 8. The peak firing temperature is 980 °C for the Au paste and 850 °C for all the other pastes.

The criteria of the conductive pastes for the electrochemically active electrodes are:

1. high purity of the fired layer;
2. well defined and reproducible electrode surface;
3. high resolution;
4. good adhesion to the substrate;
5. chemical resistance.

The conditions of high purity combined with good adhesion are especially difficult to meet. Noble-metal pastes do not adhere to the Al_2O_3 substrate without the addition of special oxides or glass. These oxides (e.g. CuO) lead to large interference in the electrochemical response. Different pastes from Dupont (DP), Thick Film Systems (TFS) and a customized paste have been characterized according to the above criteria.

For the soldering pads DP 6120 from Dupont is used. This Pd/Ag paste has good soldering ability and good adhesion to the substrate.

The paste for the passivation layer needs to have the following specification:

1. high resolution;
2. chemical resistance;
3. excellent insulating properties in liquids.

Two different pastes from Dupont (DP 5137) and Thick Film Systems (TFS 1007) have been characterized according to the above criteria.

1. DP 5137 is an overglaze paste, fired at a low temperature (500 °C). The resolution of this paste is insufficient; holes in the passivation layer of 300 μm × 300 μm cannot be realized. The chemical resistance in acids and in a 1% $FeCl_3$ solution is also low. This paste is therefore of no use for planar chemical sensors.

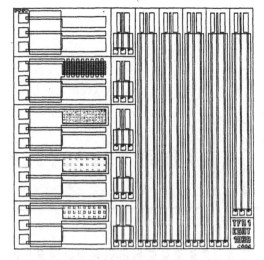

Figure 6.2 The layout of the thick-film sensor 'TFS 1'. Fifteen sensors are placed on one substrate.

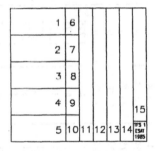

Figure 6.3 The numbering of the sensors on the 'TFS 1' layout.

2. TFS 1007 is a dielectric paste, fired at 850 °C. A high resolution can be achieved. The chemical resistance is excellent for most applications. In strong acids (pH < 3), however, this paste is affected. In order to increase the insulating properties and to decrease the pinhole density, the paste is printed twice.

6.1.2 The mask set 'TFS 1'

The layout of the thick-film sensor array 'TFS 1' is shown in figure 6.2. The design was conceived in 1985. Different types of sensors are placed on the same substrate, which measures 2 inches by 2 inches. The smallest sensor measures 10 mm by 5 mm. The geometrical characteristics of the different sensors are summarized in table 6.1 and the numbering of the sensors is indicated in figure 6.3.

Table 6.1   Geometrical characteristics of the different 'TFS 1' sensors.

sensor	dimension HxW [mm]	area WE [mm²]	# WE	area AE [mm²]	area RE [mm²]	distance WE-RE [µm]	distance AE-RE [µm]
1	20 x 10	22.5	1	28.8	9	1000	500
2	20 x 10	6.75	54	24.75	9	1000	500
3	20 x 10	6.75	75	24.75	9	1000	500
4	20 x 10	2.16	24	24.75	9	1000	500
5	20 x 10	4.32	27	24.75	9	1000	500
6	10x5	3	1	4	1.275	250	250
7	10x5	3	1	4	1.275	300	300
8	10x5	3	1	4	1.275	350	350
9	10x5	3	1	4	1.275	400	400
10	10x5	3	1	4	1.275	450	450
11	50x5	3	1	4	1.275	450	450
12	50x5	3	1	4	1.275	450	450
13	50x5	3	1	4	1.275	450	450
14	50x5	3	1	4	1.275	450	450
15	43x5	3	1	4	0.84	500	500

Sensors 1 to 5 are intended for a study on the behaviour of microelectrodes. Sensor 1 is a plain three-electrode voltammetric sensor with a large working electrode area. Sensors 2 to 5 are multi-microelectrodes. In comparison with normal electrodes, microelectrodes have a lower flow dependence and a shorter response time. The microelectrodes on sensor 2 are fabricated by printing narrow stripes of the passivation layer perpendicular to narrow stripes of the working electrode material. The microelectrodes on sensor 3 to 5 are fabricated by leaving small holes (300 μm $\times$ 300 μm or 400 μm $\times$ 400 μm) in the passivation layer. From the technological viewpoint, the first sensor is easier to realize if the resolution of the pastes is insufficient.

Sensors 6 to 10 are intended for a study on the influence of electrode spacing on the lifetime of the sensor. A narrow electrode spacing can enhance migration of Ag, for example. The Ag migration phenomenon in screen-printed circuits is well known (Ripka and Harsanyi 1985, Coleman and Winster 1981). These authors state: 'Using gaps narrower than 200 μm reliability problems can occur in thick-film circuits. If silver is present in a conductor migration will occur if moisture is in contact with the conductor. The only effective way to prevent this migration is to exclude moisture from the conductor surface.' The requirement regarding moisture cannot be met for a voltammetric sensor; migration will therefore be likely to occur. This type of sensor has the minimum dimensions achievable with standard screen-printing technology.

Sensor 11 to 15 are designed for direct 'in vitro' use. The electrode–soldering pad distance is large so that the sensor can be put directly into the solution without paying special attention to the packaging of the connection area.

With all these electrode configurations, a large variety of sensors can be made. In figure 6.4, a photograph of a completed substrate with RuO_2 electrodes is shown. For practical reasons, most of the experiments have been carried out with sensor types 1 and 11–15.

6.1.3 Packaging

Packaging of thick-film sensors is easy if the sensors are large. Due to the larger dimensions, a higher current is measured at the electrodes, so leakage currents are not that important. The most straightforward method is to solder a connector to the soldering pads and to cover this solder region with an epoxy (UHU plus, Araldite or Epo-Tek H54) or a silicone rubber (Dow Corning 3140). Alternatives are the use of heat shrinkable, plastic tubes and thick-film reel-type connectors. Several mounted thick-film sensors are shown in figure 6.5. In comparison with CMOS-compatible sensors, the packaging of thick film sensors is child's play.

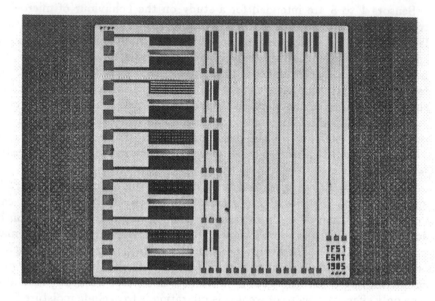

Figure 6.4 A photograph of a completed 'TFS 1' substrate with RuO_2 electrodes. The substrate measures 2 inches by 2 inches.

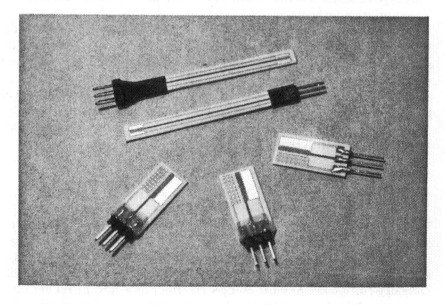

Figure 6.5 Packaged thick-film sensors with Au working and auxiliary electrodes.

6.2   ELECTROCHEMICAL EVALUATION OF THICK-FILM ELECTRODES

The electrochemical characteristics of the different pastes have been evaluated with standard techniques. The Ag electrodes were chloridated and the response to a varying Cl^- concentration was measured. The pastes for the working electrode were evaluated with cyclic voltammetry.

6.2.1   Silver electrodes

The Ag electrodes are used as Ag/AgCl reference electrodes in the three-electrode voltammetric configuration. After laser scribing and packaging, the Ag electrodes are therefore chloridated in a 1% $FeCl_3$ solution for 1 min. This procedure is the same as for the CMOS-compatible Ag/AgCl electrodes. Three different Ag pastes have been investigated for use as an Ag/AgCl electrode.

1. DP 6120 — a mixed-bonded Pd/Ag paste.

The Pd is added to this paste to decrease the migration of Ag. It is a standard paste used in the manufacture of hybrid circuits in ESAT–MICAS.

2. TFS 4055 — a fritless Ag paste.

After firing, a relatively thin film is obtained.

3. TFS 4050 — a fritless Ag paste.

It is basically the same paste as TFS 4055 but with a higher purity of the Ag film after firing.

All these three pastes have good resolution and excellent adhesion to the ceramic Al_2O_3 substrate. The DP 6120 paste is not a pure Ag paste, so it cannot be used as a reliable reference electrode. This is proved by measuring the potential at the chloridated electrodes; the difference in potential between the planar thick-film electrode and a conventional reference electrode with an internal reference electrolyte is recorded for a varying chloride ion concentration. For a good reference electrode, a Nernstian response is expected.

The response of the three pastes is shown in figure 6.6. It is clear from this figure that no good Ag/AgCl reference electrodes can be madewith the DP 6120 paste: there is a shift in the potential of 200 mV; the stability and the linearity are also insufficient. The response of the TFS 4050- and TFS 4055-based Ag/AgCl reference electrodes is Nernstian and stable. No difference has been observed between the response of TFS 4050 and TFS 4055 electrodes. The response of these thick-film Ag/AgCl electrodes is also identical to the response of the CMOS-compatible Ag/AgCl electrodes.

The ageing effect, as seen with CMOS-compatible Ag/AgCl electrodes, was also observed with the TFS samples. Initially, there is some drift, but after a few hours a potential stable within 1 mV is obtained. Because of

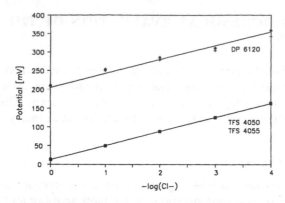

Figure 6.6 The response of three different pastes for a varying chloride ion concentration. The thick-film Ag electrodes are chloridated in a 1% $FeCl_3$ solution prior to the measurement.

the increasing surface contamination of the Ag electrode, the initial stabilization time and deviation of the potential will increase if the chloridation is not done immediately after the processing of the sensors.

It can be concluded that the electrochemical characteristics of the planar Ag/AgCl electrodes made with TFS 4050 or TFS 4055 are identical to those of classical Ag/AgCl electrodes. They have been used with success as a reference electrode in a thick-film glucose sensor.

6.2.2   Gold electrodes

The search for a good paste for the manufacture of planar thick-film Au electrodes for oxygen sensors has not been very successful. No good paste has been found by the authors. Three different pastes were evaluated.

1. DP 4119 — an oxide-bonded Au paste.

It is a standard Au paste for hybrid circuits at ESAT–MICAS. It is assumed that this paste also contains some Pd or Ag.

2. TFS 4007 — an oxide-bonded Au paste.

It contains CuO as binding a material and has a high percentage of Au in the fired layer.

3. TFS 4007 PG — a pure gold paste.

It is the same paste as the normal TFS 4007, but without CuO.

All the pastes can be screen printed with a high resolution. The adhesion to the substrate is excellent for the DP 4119 and the TFS 4007 pastes. The TFS 4007 PG paste without CuO does not adhere to the Al_2O_3 substrate and is of no use for the fabrication of planar voltammetric sensors.

In figure 6.7, the first 25 sweeps of a cyclic voltammetry experiment in PBS with a DP 4119 electrode are shown. A thick-film Ag/AgCl electrode

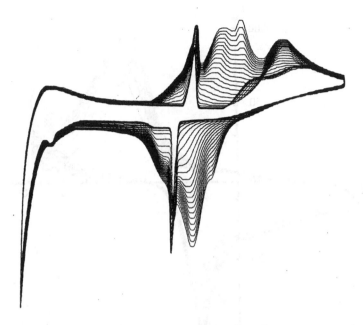

Figure 6.7   The first 25 sweeps of a cyclic voltammetry experiment in PBS saturated with air with a DP 4119 thick-film electrode.

was used as a reference electrode. In the voltammogram of the first sweep, a great variety of peaks can be observed. These peaks gradually disappear until a stable voltammogram is obtained after 25 sweeps (i.e. after 12 minutes). This indicates that a lot of impurities are present on the surface of this electrode. In the final voltammogram, the oxygen reduction wave can be distinguished. Two sharp peaks also appear on the voltammogram. These peaks are identical to those found on Ag-contaminated sputtered Pt electrodes and are a result of the reversible oxidation and reduction of Ag; this was verified with X-ray analysis. At 3 kV, Ag, SiO_2 and PbO were found in the fired layer. It is assumed that this paste also contains some Pd. Although the oxygen reduction wave is present in the voltammogram, the stability of the current is insufficient for most applications because of the impurity of the base material.

In figure 6.8, the first 25 sweeps of a cyclic voltammetry experiment in PBS with a TFS 4007 electrode are shown. A thick-film Ag/AgCl electrode is used as a reference electrode. Although initially some trace of an oxygen reduction wave is present in the voltammogram, this wave disappears after approximately ten sweeps. The only peaks that are left are due to the oxidation and reduction of Cu at the surface of the electrode; this is confirmed by X-ray analysis—a few percent of Cu is found at the surface of the electrode. Therefore, this paste is of no use for the fabrication of

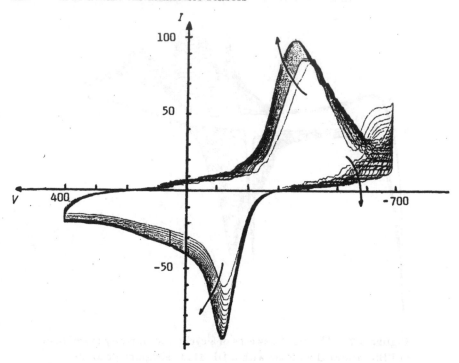

Figure 6.8   The first 25 sweeps of a cyclic voltammetry experiment in PBS saturated with air with a TFS 4007 thick-film electrode.

voltammetric sensors.

It can be concluded that the commercial pastes that we have evaluated are not suitable for the fabrication of thick-film oxygen sensors, due to impurities in the base material. These impurities, such as CuO, are however necessary for good adhesion of the Au layer to the substrate. These problems can be solved by using one of the following methods.

1. Electroplating of Au

A thin pure Au film can be plated on the existing electrodes. In the literature it is stated that plated Au films behave like bulk Au films (Westwood 1974).

2. Pure fritted Au paste

It was observed that the glass used in fritted pastes (65% PbO, 25% SiO_2, 10% Bi_2O_3) has no influence on the electrochemical response of the RuO_2 electrodes. A good composition for a custom-made Au paste seems to be 80% Au and 20% glass.

3. Organometallic pastes

There are commercial organometallic pastes available now which are practically pure Au after firing. Such pastes might result in good screen-printed Au electrodes.

4. Chemical cleaning of the electrode surface

A suitable chemical cleaning procedure is a possible solution for some of the problems related to surface impurities. It was observed that the Ag peaks in the voltammogram of the DP 4119 paste disappear if the sensor is dipped in a 0.1M HCl solution. The shape of the oxygen reduction wave, however, remained unsatisfactory.

5. Polymer-based Au pastes

Nowadays polymer-based thick-film pastes exist which are not fired at high temperatures, but only dried. As polymers are electrochemically inert and can be specified to adhere well to other surfaces, they enable good Au electrodes to be realized.

These proposed solutions have not been investigated further by the authors, since the main topic was the development of glucose sensors. The use of commercial pastes has the disadvantage that trace impurities and additives may influence the electrochemical characteristics dramatically. Since the manufacturers of thick-film materials do not reveal the composition of their pastes, it is clear that one can only know what is in a paste if one prepares the paste oneself.

6.2.3 Ruthenium dioxide electrodes

6.2.3.1 Introduction. In this work, ruthenium dioxide, a completely new material to the world of planar sensors, has been investigated as an electrode material. Conventional RuO_2 electrodes are widely studied in the literature due to their special electrochemical characteristics (Burke 1980, Trasatti and O'Grady 1981). They have an extremely low overpotential for hydrogen evolution and are used in the production of Cl_2 and as an electrode in fuel cells (Shafer *et al* 1979). Many anodic reactions which are difficult on metals proceed on RuO_2 electrodes without surface oxidation. Because of the peculiar chemisorption properties, metal-oxide electrodes are also very promising for the direct electrochemistry of redox proteins (Harmer and Hill 1985). RuO_2 is very stable up to 1200 °C and chemically resistant (Vadimski *et al* 1979). RuO_2 electrodes are usually produced by coating a Ti or Ta electrode with $RuCl_3$ and converting this $RuCl_3$ layer into RuO_2 at high temperatures in an oxygen atmosphere. With this fabrication technique, it cannot be guaranteed that the underlying Ti layer will not influence the electrochemical characteristics of the RuO_2 electrode. The formation of cracks in the oxide layer will enhance this interference. Thick-film resistive pastes are also based on conducting metal oxides: RuO_2 and $Bi_2Ru_2O_7$. These pastes are commercially available in a broad range of resistivity and composition. In this work the existence of these conducting metal oxide pastes and the interest in metal-oxide electrodes are combined into a study of planar, low-cost RuO_2 electrodes.

6.2.3.2 Preparation of thick-film RuO₂ electrodes. After an initial screening, two different pastes have been characterized for use as a planar electrode.

1. DP 8011 — a standard resistive paste produced by Dupont.

After firing, the resistive layer contains between 30% and 40% RuO_2. The porosity of the layer is 3% to 4%. The remaining part is glass; it is assumed that this glass has the following composition: 65% PbO, 25% SiO_2, 10% Bi_2O_3. The sheet resistivity is 10 Ω/square.

2. PS 8 — an experimental paste made by S Paszczynski for pressure sensor applications.

After firing, the resistive layer contains 50% $Bi_2Ru_2O_7$. The porosity of the layer is high (20% to 25%). The remaining part is glass, with the following composition: 65% PbO, 25% SiO_2, 10% Bi_2O_3. The sheet resistivity is 10 Ω/square.

These pastes have been evaluated using cyclic voltammetry in PBS. As a measure for the quality of the electrode, the sensitivity to the redox system $[Fe(CN)_6]^{3-}/[Fe(CN)_6]^{4-}$ was measured.

6.2.3.3 Electrochemical characterisation of RuO₂ electrodes. In figure 6.9, the cyclic voltammogram of a thick-film RuO_2 electrode in PBS is shown. As a reference electrode and an auxiliary electrode, the planar electrodes on the device itself have been used: the reference electrode is an Ag/AgCl electrode; the auxiliary electrode is a RuO_2 electrode. The voltammogram shows that the current is non-faradaic in the potential region between 750 mV and −450 mV. No chemical reaction or adsorption effects of hydrogen and oxygen are seen. The residual current is purely capacitive and is the result of the charging and discharging of the electrochemical double layer. Considering the scan rate and the electrode area, a double-layer capacitance of 150 μF cm^{-2} is derived.

Cyclic voltammograms for the redox system $[Fe(CN)_6)]^{3-}/[Fe(CN)_6]^{4-}$ have been recorded for different concentrations. In figure 6.10, the cyclic voltammograms for a varying concentration (0 to 9 × 10^{-4} M) are shown. Two peaks, typical for the redox system $[Fe(CN)_6)]^{3-}/[Fe(CN)_6]^{4-}$, appear on the voltammogram. The current at the peak maxima and at the anodic (600 mV) and cathodic (−250 mV) plateau are plotted in figure 6.11 as a function of the concentration. A linear response is obtained at the four different points. The highest sensitivity is reached at the anodic peak.

The influence of the scan rate on the electrode response has been verified by recording the cyclic voltammogram for different scan rates in PBS, PBS (10^{-4} M $[Fe(CN)_6]^{4-}$) and PBS (10^{-3} M $[Fe(CN)_6]^{4-}$). The results are summarized in figure 6.12, which plots the residual capacitive current and the faradaic current at the anodic peak as a function of the scan rate. The faradaic current is obtained by subtracting the residual current from

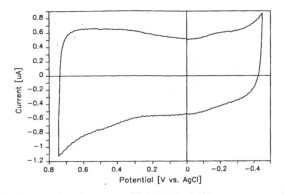

Figure 6.9 Cyclic voltammogram of a thick-film RuO_2 electrode in PBS (area $WE = 3$ mm^2, scan rate $= 100$ mV s^{-1}).

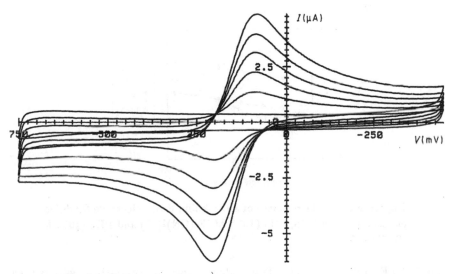

Figure 6.10 Cyclic voltammograms of a thick-film RuO_2 electrode in PBS for a varying $[Fe(CN)_6]^{4-}$ concentration (0 to 9×10^{-4} M).

the measured current. The current in PBS (i.e. the capacitive current) is linearly proportional to the scan rate. This is logical, as this current is the result of the charging and discharging of the electrochemical double layer. The current, i_c, through this capacitor, C_d, is defined as (formula (2.26)):

$$i_c = C_d S. \tag{6.1}$$

This is exactly the response that is measured for the thick-film RuO_2 electrodes. From this experiment, the double-layer capacitance can be calculated. Taking into account the electrode area, a double-layer capacitance

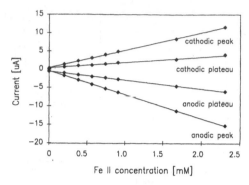

Figure 6.11   The current at the peak maxima and at the anodic (600 mV) and cathodic (−250 mV) plateau as a function of the concentration for a thick-film RuO_2 electrode in PBS for the FeII–FeIII redox system.

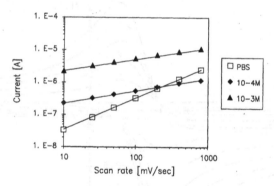

Figure 6.12   The response of a thick-film RuO_2 electrode for different scan rates in PBS, PBS (10^{-4} M $[Fe(CN)_6]^{4-}$) and PBS (10^{-3} M $[Fe(CN)_6]^{4-}$).

of 100 μF cm^{-2} is found. This value is low in comparison with values reported in Burke (1980) for RuO_2-coated Ti electrodes.

According to the electrochemical theory (formula (2.36)), the faradaic current is proportional to the square root of the scan rate if the electrode process is completely reversible. The faradaic current, i_f can be expressed as:

$$i_f = KA[FeII]S^{\alpha}$$

where K is a constant, A is the working electrode area (mm^2), [FeII] is the $[Fe(CN)_6]^{4-}$ concentration (M), S is the scan rate (mV s^{-1}) and α is a constant.

The value of the constant α is 0.5 for a reversible process. For the thick-film RuO_2 electrodes, a value of 0.37 is found. This value is equal to the gradient of the straight lines in figure 6.12. At any scan rate, the

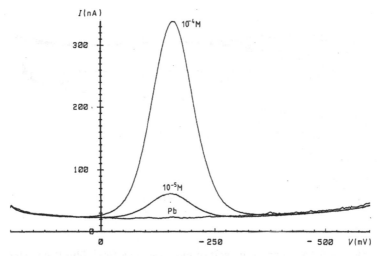

Figure 6.13   Square-wave voltammogram of a RuO_2 thick-film electrode in PBS, PBS (10^{-5} M $[Fe(CN)_6]^{4-}$) and PBS (10^{-4} M $[Fe(CN)_6]^{4-}$).

faradaic current increases by a factor of ten when increasing the concentration from 10^{-4} to 10^{-3}. This means that the response of the sensor is linear, independent of the scan rate. From this measurement, it can also be concluded that in order to increase the sensitivity of the sensor, the scan rate has to be low. If the scan rate is decreased, the residual capacitive current drops faster than the faradaic current and so the sensitivity increases. From the faradaic current and the value of the diffusion coefficient of $[Fe(CN)_6]^{4-}$, the active electrode area can be calculated. This has been carried out in Claes and Maene (1987); they came to the conclusion that the active electrode area is only 5% of the geometrical electrode area. This can be explained by the knowledge that more than 50% of the fired layer is glass. If this glass covers the RuO_2 particles, these particles become electrochemically inactive. Therefore, only 5% of the electrode surface is not covered with glass. This indicates that the properties of RuO_2 electrodes can be improved by increasing the quantity of RuO_2 particles in the paste. An alternative measuring technique (square-wave voltammetry) was used to improve the sensitivity. With this technique a concentration of 10^{-5} M $[Fe(CN)_6]^{4-}$ can easily be detected with good resolution, as can be seen in figure 6.13, where the square-wave voltammogram of a thick-film RuO_2 electrode is shown for PBS, PBS 10^{-5} M $[Fe(CN)_6]^{4-}$ and PBS 10^{-4} M $[Fe(CN)_6]^{4-}$. With square-wave voltammetry, the residual current is cancelled and the peaks are clearly distinguished. It is also an excellent technique for the evaluation of electrolytes with different electrochemically active species. However, it is a complicated measuring technique and it

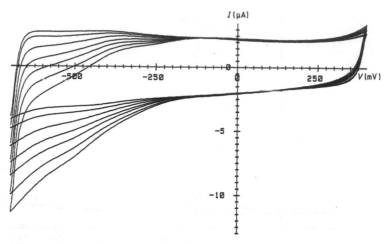

Figure 6.14   The cyclic voltammograms of a thick-film RuO_2 electrode for a varying H_2O_2 concentration (0–10^{-3} M) in PBS.

is therefore not suitable for 'in vivo' measurements. Since the interesting potential range of RuO_2 electrodes is situated in the negative potential region, the response to H_2O_2 has been evaluated. An excellent H_2O_2 response is essential for the construction of H_2O_2-based biosensors, such as glucose sensors.

6.2.3.4   Applicability of RuO_2 electrodes as H_2O_2 detectors.   The electrochemical response of RuO_2 electrodes in PBS with H_2O_2 has been investigated with cyclic voltammetry, linear-sweep voltammetry and amperometry (Claes and Maene 1987). With this study, the applicability of RuO_2 electrodes as an H_2O_2 detector for a planar glucose sensor was verified. For all these measurements, the planar auxiliary and reference electrodes on the device itself are used. In figure 6.14, the cyclic voltammograms for a varying H_2O_2 concentration (0–10^{-3} M) in PBS are shown. A faradaic current which depends on the H_2O_2 concentration is found at potentials above 200 mV. The shape of the current wave indicates that the electrochemical process is irreversible. The H_2O_2 in the solution is oxidized according to the following reaction:

$$H_2O_2 \rightarrow O_2 + 2H^+ + 2e^-.$$

In figure 6.15, the current response at four different potentials (400, 500, 600 and 700 mV versus AgCl) is depicted for a varying H_2O_2 concentration (0–10^{-3} M) in PBS. An excellent linearity is observed for all these potentials. The best sensitivity is obtained at a potential of 700 mV versus

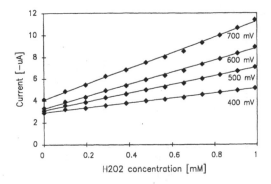

Figure 6.15   The current response of a thick-film RuO_2 electrode at four different potentials (400, 500, 600 and 700 mV versus AgCl) for a varying H_2O_2 concentration (0–10^{-3} M) in PBS.

AgCl. The lowest concentration that can be measured with an acceptable faradaic to residual current ratio is 10^{-4} M. This corresponds to an adequate sensitivity for a H_2O_2-based glucose sensor.

The influence of the scan rate on the H_2O_2 response was also examined. The current at 700 mV versus AgCl in the cyclic voltammogram was recorded as a function of the scan rate for PBS, PBS (10^{-4} M H_2O_2) and PBS (10^{-3} M H_2O_2). The results are summarized in figure 6.16, where the residual current and the faradaic currents are plotted as functions of the scan rate on a logarithmic scale; the faradaic current is obtained by subtracting the residual current from the measured current. As already stated in section 6.2.3, the capacitive current is linear as a function of the scan rate (formula (6.1)) and the faradaic current is described by formula (6.2). If compared to the response for $[Fe(CN)_6]^{4-}$, the faradaic current is practically independent of the scan rate. This proves that the oxidation of H_2O_2 at the RuO_2 electrodes is an irreversible process. It also implies that a low scan rate will improve the sensitivity of the RuO_2 electrodes to H_2O_2.

For quantitative measurements, linear-sweep voltammetry is used instead of cyclic voltammetry. With linear-sweep voltammetry, similar results to those with cyclic voltammetry have been obtained. In figure 6.17, the response at four different potentials (400, 500, 600 and 700 mV versus AgCl) is depicted for a varying H_2O_2 concentration (0–10^{-3} M) in PBS. The voltammogram was recorded between 0 mV and 700 mV at a scan rate equal to 100 mV s^{-1}. A rest potential equal to 0 mV was applied to the sensor for 7 seconds prior to each sweep. An excellent linearity is noted. The highest sensitivity is also obtained at a potential of 700 mV versus AgCl. The stability and reproducibility of the measurement are excellent; a maximum deviation of only 2% was observed under varying experimental conditions (Claes and Maene 1987). Amperometry can also be used for the detection of H_2O_2; this technique is very fast and does

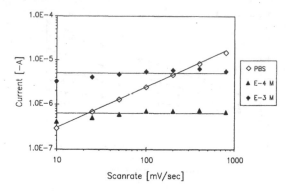

Figure 6.16 The response of a thick-film RuO_2 electrode for different scan rates in PBS, PBS (10^{-4} M H_2O_2) and PBS (10^{-3} M H_2O_2).

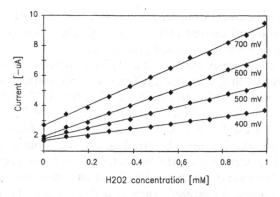

Figure 6.17 The linear-sweep voltammetry current response of a thick-film RuO_2 electrode at four different potentials (400, 500, 600 and 700 mV versus AgCl) for a varying H_2O_2 concentration ($0-10^{-3}$ M) in PBS.

not require complicated equipment. For amperometry of H_2O_2 at RuO_2 electrodes, the potential is fixed at 700 mV versus AgCl. The response time is very fast, as can be seen in figure 6.18. In this experiment, 100 μl of a concentrated H_2O_2 (10^{-2} M) solution is added sequentially to 10 ml of a stirred PBS solution; a high sensitivity and a linear calibration curve are obtained (see figure 6.19). The stability of the residual current and the faradaic current are appropriate for the realization of a H_2O_2-based glucose sensor, as will be discussed in 6.3.

6.2.3.5 Conclusion. The new electrode material, RuO_2, has interesting electrochemical properties. Planar RuO_2 electrodes have been fabricated using commercially available pastes. These thick-film RuO_2 electrodes

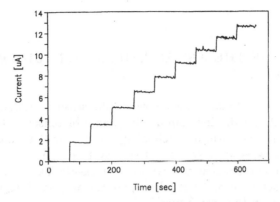

Figure 6.18   Amperometric response of a thick-film RuO_2 electrode to step-wise changes of the H_2O_2 concentration in a stirred PBS solution.

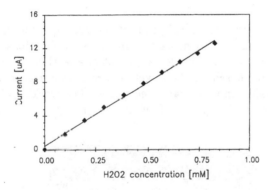

Figure 6.19   Calibration curve obtained from the measurement in figure 6.18.

behave as ideal electrodes. The residual current is purely capacitive in PBS; no interfering reactions take place in the potential region between −400 mV and 700 mV versus AgCl. The response to the redox system $[Fe(CN)6]^{3-}/[Fe(CN)_6]^{4-}$ follows the basic theory of electrochemistry (see section 2.3.1). These screen-printed thick-film electrodes do not suffer from interference of the underlying material, which is the case for the classical RuO_2-coated Ti electrodes. Thick-film RuO_2 electrodes are especially suitable for anodic applications and are therefore an alternative to the less stable Pt electrodes. A promising application of these thick-film RuO_2 electrodes is their use as H_2O_2 detectors in planar biosensors; the response to H_2O_2 is linear, reproducible and stable. In combination with a suitable enzymatic membrane, a completely new breed of biosensors could emerge. As a test case, a thick-film glucose sensor has been developed for 'in vitro' applications. Results on this type of glucose sensor are described in the next section.

6.3   STATE OF THE ART IN THICK-FILM BIOSENSOR RESEARCH

In this section several realizations of thick-film sensors are discussed. Taking into account that the first commercial planar biosensor on the market is fabricated with screen-printing technology, it is clear that thick-film technology based sensors are a cost-effective way to produce biosensors. It is believed by the authors that several other thick-film based biosensors will enter the market soon and will also open the way for integrated-circuit based biosensors in the near future.

6.3.1   Thick-film glucose sensors developed at K U Leuven, ESAT–MICAS

Thick-film technology is the ideal production technique for the fabrication of moderate volumes of biosensors for 'in vitro' applications. This statement has been validated by the development of a working prototype thick-film glucose sensor, based on the novel electrode material RuO_2, which is used for the working and auxiliary electrodes. The working electrode acts as an H_2O_2 detector. The process flow and the results are discussed in this section.

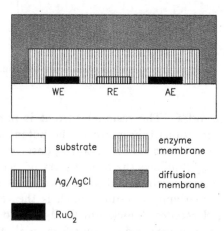

Figure 6.20   The cross section of a thick-film glucose sensor with glucose oxidase and diffusion membrane.

The cross section of a thick-film glucose sensor is presented in figure 6.20. The screen-printed three-electrode system 'TFS 1' described in the previous section is used as a base for the glucose sensor. On top of the three-electrode system, a glucose oxidase membrane and a diffusion membrane is applied. The working and auxiliary electrodes are made in RuO_2 paste (DP 8011) and the Ag/AgCl reference electrode is made in TFS 4050 or TFS 4055. The substrate is a common Al_2O_3 substrate. In the enzymatic glucose oxidase membrane, the glucose is converted into H_2O_2. As a matrix, bovine serum albumin has been chosen because of the well documented preparation of this type of membrane. The enzyme glucose oxidase is covalently bound to the matrix using glutaraldehyde. Special attention has been paid to the adhesion of the membrane to the electrode surface and the substrate. To increase the linear range, a diffusion membrane is applied over the enzymatic membrane. Good results have been obtained with cellulose acetate and polyurethane membranes. For the evaluation of the glucose sensors, linear-sweep voltammetry as well as amperometry were used. The most straightforward method, amperometry, gave excellent results.

6.3.1.1 Preparation of the enzyme membrane.   For the preparation of the thick-film glucose sensor, the following process flow has been derived.

1. Preparation of a three-electrode thick-film H_2O_2 detector
The RuO_2 working and auxiliary electrodes are printed with the DP 8011 paste. The Ag/AgCl reference electrode is made in TFS 4050 or TFS 4055 (see section 6.1.1).

2. Cleaning of the sensor in acetone

3. Silanization of the electrode surface
The sensor is dipped in a 1% γ-aminopropyltriethoxysilane in spectrophotometric-grade methanol solution with 4% H_2O for 60 seconds. The sensor is blown dry with nitrogen and baked for 5 minutes in air at 110 °C. In this way, in principle a monolayer of γ-aminopropyltriethoxysilane is formed at the surface of the electrode and the substrate. A covalent bond is formed between the γ-aminopropyltriethoxysilane and the hydroxyl (OH) groups at the oxide surface.

4. Fixation of the bifunctional reagent glutaraldehyde
By immersion in a 5% glutaraldehyde in H_2O solution, the bifunctional reagent reacts with the γ-aminopropyltriethoxysilane layer and a covalent bond is formed. The sensor is rinsed in H_2O and blown dry with nitrogen.

5. Preparation of the enzyme membrane solution
A typical enzyme solution consists of the following components:
- 70 mg bovine serum albumin in 1 ml PBS,
- 2.2 mg glucose oxidase in 0.1 ml PBS (activity GOD= 250 units mg^{-1}),
- 0.2 ml glutaraldehyde 5% solution.
The bovine serum albumin solution is mixed with the glucose oxidase solution and the glutaraldehyde solution is then added.

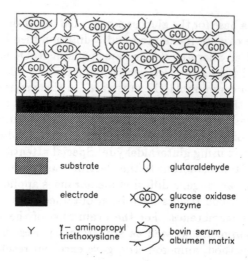

substrate   glutaraldehyde

electrode   glucose oxidase enzyme

γ− aminopropyl triethoxysilane   bovin serum albumen matrix

Figure 6.21   Cross-linking mechanism of the glucose oxidase enzyme in the bovine serum albumin matrix. A covalent bond is formed between the matrix and the substrate.

6. Casting of the glucose oxidase membrane

A well determined quantity of the enzyme membrane solution (typically 45 μl for a type 1 sensor) is applied on the sensor tip immediately after the addition of the glutaraldehyde solution.

7. Polymerization of the glucose oxidase membrane

A gel is formed within 15 minutes after addition of the glutaraldehyde solution. A covalent bond is formed between the proteins (BSA and GOD) with the bifunctional reagent glutaraldehyde as cross-linker. In this way a gel matrix is formed by covalent coupling of the enzyme to the bovine serum albumin matrix. The proteins also react with the glutaraldehyde on the surface of the electrode, so a covalent coupling of the glucose oxidase membrane to the sensor surface is achieved. After polymerization at room temperature for 1 hour, the sensor is lyophylized.

It has been examined in what way the different steps in the preparation of the sensor influence the electrochemical response of the RuO_2 electrode to H_2O_2. Although a small influence has been measured, the H_2O_2 response is still sufficient for the manufacture of glucose sensors (Claes and Maene 1987). With this process sequence, a covalently coupled glucose oxidase membrane based on a bovine serum albumin matrix is realized with excellent adhesion to the substrate. The activity of the glucose oxidase enzyme in the membrane after the complete process was rated at 10% of the initial activity. Although this decrease in activity is large, it is inevitably due to the covalent coupling of the enzyme. During this process, several active sites at the enzymes are blocked. A chain of covalent bonds holds the membrane onto the substrate. Covalent coupling of the membrane to the electrode surface is a necessity for the realization of planar sensors, as no mechanical attachment can be used. An overview of the different bonds can be seen in figure 6.21.

This method, based on casting, has been transformed into a dip-coating method in order to decrease the membrane thickness and increase the uniformity. The process sequence remains identical except that the membrane is applied on the sensor by dip coating. Excellent glucose sensors have been made with both methods (see section 6.3.1). However, both processes are based on manual techniques. For the mass-production of thick-film sensors, the method detailed above has to be converted into a mass-production technique. By modifying the enzyme solution so that a suitable viscosity and thixotropy is reached, the enzyme layer can be screen printed. In order to achieve an optimum production technique, the polymerization process also has to be modified. Using the present polymerization process, the screen has to be cleaned after each print step. By increasing the gel formation time to several hours or by switching to a photochemical cross-linking method, this problem can be circumvented.

6.3.1.2 Results. The glucose sensors described in the previous section have been evaluated using linear-sweep voltammetry. As a typical scan rate, 10 mV s^{-1} was used with a 20 second pre-polarization at 100 mV. In figure 6.22, the linear-sweep voltammograms of a cast thick-film glucose sensor in PBS with different glucose concentrations (2.5, 5, 10, 15, 20 and 25 mg dl^{-1}) are shown. In the potential range between 250 mV and 700 mV an increase in the anodic current is found for an increasing glucose concentration. The response curves (figure 6.23) show a linear response up to 25 mg dl^{-1}. Above this concentration a saturation in the response is observed. This is due to oxygen deficiency in the glucose oxidase membrane; under these conditions the enzymatic reaction cannot occur and the current saturates. The limit in the linear range of 25 mg dl^{-1} corresponds well with values reported by other authors for covalently bonded bovine serum albumin based membranes. An extended optimization has been carried out concerning the membrane composition and thickness (Claes and Maene 1987). It was concluded that a low enzyme concentration and a thin membrane gave the most interesting results. The response of one of the best sensors that has been realized is shown in figure 6.24. On this 'type 1' sensor, 0.06 mg glucose oxidase is applied as a 45 μl enzyme solution using the method described above. A linear range up to 30 mg dl^{-1}, an excellent linearity and a good reproducibility have been observed. The sensitivity to glucose of this sensor at 650 mV is 1.24 mA M^{-1}.

The response time of the cast sensors is approximately 1 minute. The operational lifetime is more than one month if measured daily and stored in PBS at 4 °C when not in use (see figure 6.25). A gradual increase of the

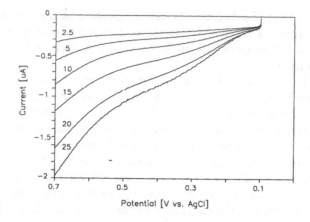

Figure 6.22   Linear-sweep voltammogram of a cast thick-film glucose sensor in PBS with different glucose concentrations (2.5, 5, 10, 15, 20 and 25 mg dl^{-1}).

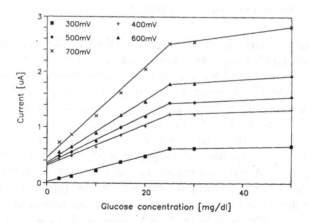

Figure 6.23   Response curve of a cast thick-film glucose sensor in PBS for a varying glucose concentration.

current is seen as a function of time. This phenomenon can be explained by a decrease in the membrane thickness as a function of time or by dissolution of the glass at the surface of the RuO$_2$ electrode. The lifetime of this sensor is more than sufficient; thick-film sensors are intended as disposable sensors. For long-term use it is important to note that the activity of the sensor does not decreases. This means that the sensor remains sensitive to glucose; a daily calibration before use is necessary, however, to compensate the drift of the device.

From these measurements it can be concluded that a working glucose sensor with excellent lifetime, response time and sensitivity has been real-

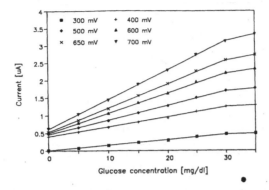

Figure 6.24 Response curve of a cast thick-film glucose sensor in PBS for a varying glucose concentration. The sensitivity to glucose of this sensor at 650 mV is 1.24 mA M^{-1}.

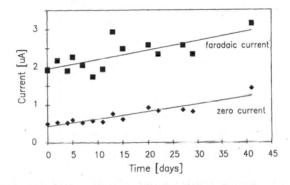

Figure 6.25 Operational lifetime test—response of a cast thick-film glucose sensor in PBS and PBS with 20 mg dl^{-1} glucose during 41 days. The residual current and the faradaic current are plotted as a function of time.

ized. The limited linear range is insufficient, however, for the evaluation of undiluted clinical samples. A further enhancement of the sensor is therefore necessary.

It was also seen that dip coating of the sensor in the enzyme solution instead of casting gave better results. At the same time amperometry was chosen for the evaluation of the sensor. A glucose sensor can be tested more quickly with amperometry by injection of small amounts of a concentrated glucose stock solution into stirred PBS. The same approach, dip coating and amperometry, has been used for the evaluation of diffusion membranes. Two types of diffusion membranes are investigated; cellulose acetate and polyurethane. For the preparation of cellulose acetate membranes, 250 mg cellulose acetate is dissolved in 25 ml acetone. The sensor is dipped into

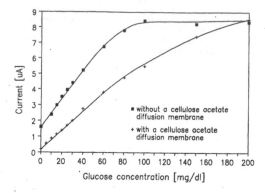

Figure 6.26   Amperometric response of a dip-coated thick-film glucose sensor with and without a cellulose acetate diffusion membrane.

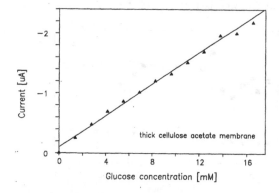

Figure 6.27   Amperometric response of a dip-coated thick-film glucose sensor with a thick cellulose acetate diffusion membrane.

this solution and dried. A typical response of such a sensor is shown in figure 6.26. It is seen that the cellulose acetate membrane extends the linear region up to 100 mg dl^{-1}. By increasing the thickness of the diffusion membrane, the linear range can even be extended up to 300 mg dl^{-1} (figure 6.27). A loss in sensitivity, however, is inevitable. It is known from the literature that with a cellulose acetate diffusion membrane, excellent glucose sensors can be made with high linearity. However, due to the porous fibre structure, blood clotting occurs very easily on the sensor resulting in deterioration of the response. During short-term or single measurements, the clotting will not influence the response of the sensor. It implies, however, that glucose sensors with a cellulose acetate diffusion membrane can be used only for long-term measurement of glucose levels in plasma, not whole blood. For these experiments, a polyurethane or Nafion membrane is necessary.

6.3.1.3 Conclusion. Working thick-film glucose sensors have been realized. A 'TFS 1' thick film sensor with an Ag/AgCl reference electrode and RuO_2 working and auxiliary electrodes is used as an H_2O_2 detection unit. The enzyme membrane consists of a classical bovine serum albumin matrix with covalently bonded glucose oxidase. This membrane is covalently bonded to the substrate with silanizing agents and glutaraldehyde, resulting in excellent adhesion and a long lifetime of the enzyme membrane; an operational lifetime of more than one month has been recorded. A response time of 1 minute and a linear range up to 30 mg dl^{-1} has been realized without a diffusion membrane. With a cellulose acetate diffusion membrane, the linear range can be extended up to 300 mg dl^{-1}. Until now the membranes have been applied on the sensor by dip coating and casting. For the realization of cheap, disposable thick-film sensors, this manual process has to be converted into a screen-printing process.

6.3.2 Thick-film oxygen sensors

The realization of planar thick-film oxygen sensors was not of primary concern to the authors. Therefore, no special attempt has been made to development a good Au paste or to modify the existing layout. Since the auxiliary electrode of a biomedical oxygen sensor cannot be made in Au because of the chloridation of the electrode, the existing mask set has to be changed and extended. A separate mask and screen are necessary for the auxiliary electrode and the working electrode. The auxiliary electrode can then be made in a different material such as Pt or RuO_2. Other authors, however, claim that they have fabricated a thick-film oxygen sensor for medical applications with an Au auxiliary electrode (Karagounis *et al* 1986). This is not good practice as this way the lifetime of the sensor is reduced and migration problems will occur.

The dissolved-oxygen sensor which is most evolved is described in Pace *et al* (1985). This two-electrode sensor consists of several multilayers (see figure 6.28). All layers are screen printed with specially developed pastes (Pace 1986). An Ag paste is used for a reference electrode and the working electrode is made in Pt. On top of the active electrode area, a hydrogel layer is screen-printed consisting of a PVA–KCl mixture and an oxygen diffusion membrane consisting of poly(styrene/acrylonitrile). The entire sensor is overlayed with a silicone rubber top layer to make it blood-compatible. The sensor is rather large in comparison to sensors fabricated with integrated-circuit technology (25.4 mm by 25.4 mm). Measurements indicate that a thick-film pO_2 sensor has effectively been designed for medical diagnostic applications. In order to overcome the limitations of the amperometric approach, calibration of the sensor with ambient air combined with a voltage-pulse technique was found to be necessary (Pace *et al* 1985).

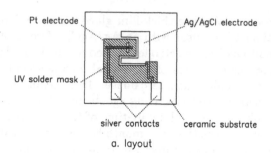

a. layout

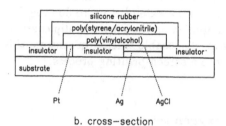

b. cross—section

Figure 6.28 Screen-printed oxygen electrode (adapted from Pace *et al* (1985)).

6.3.3   The 'ExacTech' approach: a commercial example of a thick-film biosensor

Although initially marketed as a 'Blood Glucose Test Strip' and not as a 'Blood Glucose Biosensor', the ExacTech blood glucose monitoring system is based on a planar screen-printed biosensor. It is the first example of a planar biosensor that has entered the commercial market. In papers and leaflets no details are revealed about the production process, but after close examination it is clear that this sensor has been fabricated using thick-film or screen-printing technology. Only in recent publications is it stated that the ExacTech sensors are fabricated with a screen-printing process (Turner 1990). The new device was launched on the market in the USA in 1987 by Baxter Travenol (Turner 1989). The technology of this biosensor is based on a ferrocene-mediated amperometric glucose oxidase configuration. This technology was invented jointly by researchers at Cranfield Institute of Technology and Oxford University (Cass *et al* 1984). A patent on this sensor for components of a liquid mixture was assigned to Genetics International, now MediSense (Higgins *et al* 1985). The complete system consists of disposable glucose sensor strips and a pen-like potentiostat and display unit. The system is produced by MediSense at two locations in Boston and Abingdon, England; the electronic pen is assembled in Boston while the disposable strip is manufactured in England. The system was initially distributed world-wide by Baxter Travenol, a large multinational in medi-

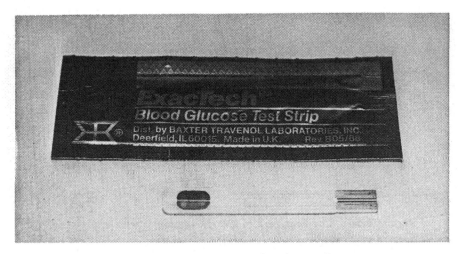

Figure 6.29 The ExacTech blood glucose test strip shown in combination with its package.

cal materials. Nowadays the device is distributed world-wide by MediSense itself. In figure 6.29, a photograph of a disposable sensor and the corresponding package is shown. The sensor measures 7 mm by 45 mm. The two-electrode sensor consists of a plastic carrier on which different layers are screen printed. The reference electrode is made in a silver-based paste; the working electrode is made in a carbon-based paste. Glucose oxidase enzyme and ferrocene are also incorporated on the sensor in a dry form. The active sensor area and the connection electrodes are delineated with a white polymer covering layer. A special surface coating and a polymer mesh allow a drop of blood to flow uniformly over the active electrode area.

The blood glucose meter incorporates a microchip and an LCD display, as shown in figure 6.30. The pen is 135 mm long and has a diameter of 10.5 mm; it weights only 30 g. The non-replaceable batteries provide enough energy for two years or approximately 4000 measurements. The systems gives a response in 30 seconds. Clinical tests have shown patient acceptance and sufficient accuracy of this system (Matthews *et al* 1987). Also, a new meter, named Companion, has recently been launched by MediSense. This meter is the size of a credit card. The main advantage is the larger display so that people who have eye problems (a lot of diabetics suffer from poor eyesight) can use the system. The main advantage of this system is that, in comparison with classical glucose meters based on densitometry of colour reagent strips, no blood-contaminated strips enters the meter. This blood-stained end can contaminate the optical read-out system of classical meters and result in inaccurate measurements. With the ExacTech system, a fast reading with good accuracy is obtained. Other biosensors are in development by Medisense. According to Clinica (1990),

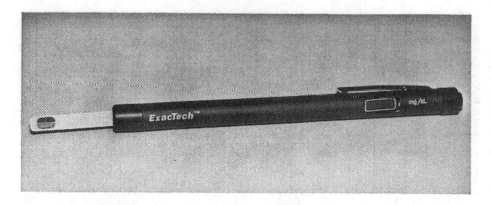

Figure 6.30   The ExacTech blood glucose meter with a test strip inserted ready for use.

FDA approval has been sought for six biosensor-based diagnostics tests—the Satellite glucose system, cholesterol, potassium, paracetomol, alcohol and a second-generation glucose test. It is also stated that the new Satellite system includes facilities to compensate for interference in the blood, a built-in temperature sensor and an on-board printer. A second-generation glucose sensor has reached the market (MediSense 1991, McKinnon *et al* 1991)—a third electrode is now incorporated in the system. This electrode is a second working electrode without immobilized GOD enzyme: it measures all non-glucose related currents, allowing compensation for any interference in the blood sample. The new system is automatically activated when blood is applied on the sensor tip. The name ExacTech has been abandoned and it is now marketed as the MediSense system. Three different electronic meters exist: the new Satellite G and, for self-monitoring, the MediSense Pen 2 and Companion 2. In the new leaflets, the unique advantages of biosensor technology are mentioned explicitly.

Therefore, a lot of new and commercially interesting biosensors based on thick-film technology can be expected in the very near future. A good example of how new sensors based on good working technology can easily be developed is the screen-printed sensor for detecting insecticides and nerve gases described in Kulys and D'Costa (1991). A lot of similarities can be seen between this sensor and the ExacTech device. Both systems were developed at the Cranfield Institute of technology.

6.4   CONCLUSION

In this chapter it is proved that with thick-film technology, planar voltammetric sensors can be produced. A working thick-film glucose sensor realized by the authors and a commercial blood glucose test strip based on a ferrocene biosensor manufactured by MediSense is discussed. The sensor developed by the authors is suitable for 'in vitro' determination of glucose levels. Excellent linearity, response time, operational lifetime and reproducibility have been obtained. The realization of a planar thick-film oxygen sensor was also examined by the authors. However, this was not successful due to the lack of a reliable thick-film Au electrode. It is possible, however, to make a thick-film oxygen sensor as described in Pace *et al* (1985). It can be concluded that with thick-film technology, reliable voltammetric biosensors can be made.

CHAPTER 7

CONCLUDING REMARKS

More and more people are becoming interested in sensors and more and more research and development is being devoted to sensors. Indeed sensors have become so attractive mainly because they have become a necessity in control systems. They are already in use in automobiles, in production control systems, in medicine, etc. They are an essential part of microsystems, which tend to integrate the sensor with the interface electronics and the microprocessor all on one chip. The sensor may very well be the most difficult component to integrate, however. Enhanced research efforts are thus required.

A second reason why sensors are so important is the expanding market. In 1989 Toyota produced 3 millions sensors, increasing at 30% per year. Biosensor market information can be found over all orders of magnitudes, depending on the source. An optimistic source expects that the biosensor market for medical diagnostics in the USA will reach a value of 10^9 by the year 2000. The total biosensor market is expected to reach 2×10^9 in the USA. More realistic sources speak about a total market of 700×10^6 for the whole world by the year 2000. These figures do not require more explanation.

A third reason why sensors are so attractive is that they require multidisciplinary research. They involve materials science, microelectronics, chemistry, micromechanics, all to a large extent. Many aspects do not even belong to any one category: packaging, for example, is such a hybrid; enzyme deposition on top of a chip is another. Many new problems can thus be defined and many new solutions are required. Moreover, universities are very well suited to play an important role because they can easily take up new challenges in multidisciplinary research.

In the area of biosensors, the discipline of microelectrochemistry has become more and more important the last few years. It combines electrochemistry with microtechniques such as planarization, both monolithic and in thick film, to result in devices such as monolithic and thick-film glucose sensors, urea sensors, dissolved oxygen sensors etc, based on voltammetry, potentiometry and conductometry. The fabrication of these devices is the main subject of this book.

This book starts with a general introduction on the definition of biosensors. It describes what biosensors are and shows that glucose sensors are probably still the most important biosensors. The glucose sensor is therefore used as a generic sensor throughout the book.

The second chapter deals with elementary electrochemistry. Techniques such as potentiometry, voltammetry and conductometry are defined. It is written for engineers but is kept as simple as possible in order to make the elementary principles understoodandable. The third chapter focuses especially on measurement principles. A large range of measurement techniques is always available. The selection of the correct technique is determined by the application.

In chapter 4, planar technologies for the realization of sensors are discussed. Both integrated circuit and thick-film techniques are included. These are vital since they allow mass production at low cost, leading to disposable sensors. Some of the techniques that are described in the second part of this chapter are specific to sensors. Examples are lift-off, micromachining and chemical membrane deposition. These special technologies are the most difficult, as existing techniques can not always be used. Therefore, new technology has to be developed.

Several case studies are discussed in chapter 5, starting with CMOS-compatible electrochemical sensors using platinum, gold and silver electrodes. Special attention is paid to sensors fabricated in the ESAT–MICAS group of the Faculty of Applied Sciences of the Katholieke Universiteit Leuven (ESAT stands for *Electronics, Systems, Automation and Technology*; MICAS stands for *Medical & Integrated Circuits and Sensors*). Other realizations are discussed as well, however.

These sensors are especially suited for in vivo applications. Due to high complexity and biocompatibility problems, no commercial devices are available on the market yet. It is, however, expected that after planar biosensors based on thick film technology have opened up the market, miniature sensors based on integrated circuit technology will emerge.

Complementary thick-film sensors are discussed in chapter 6. They have the advantage that they are relatively cheap for smaller quantities. On the other hand they are less reproducible. However, they have given rise to the first commercial biosensors. Thick-film sensors are especially suited for in vitro sensing. Thick-film or screen-printing technology therefore becomes more and more important for the realization of devices for the 'over the counter' diagnostics market. At this moment, several companies are following the route opened up by the ExacTech Medisense glucose sensor and are developing screen-printed devices for general use by doctors, nurses and patients.

Finally, the last chapter summarizes the most important conclusions.

So, it can be concluded that in this book the important steps towards the realization of cheap, disposable biosensors are explained. The contribution of the authors is mainly in the development of a universal mass-production technique for microelectrochemical biosensors. By using the same techniques as for integrated circuits, it can be expected that planar, integrated and smart sensors will show the same market growth rates as observed in the microelectronic world. This trend already exists in the mechanical sensor world; nowadays, millions of silicon-based pressures sensors are marketed at a continuously declining price (Novasensor 1987, IC Sensors 1987). Silicon-based accelerometers also look promising.

For chemical sensors, this trend and commercial success has not yet started. This can be related to the enormous variety of chemical concentrations that have to be measured. For pressure sensors, for example, there exists only 'one pressure'; this implies that the same design can be used for practically all applications—only the environment (i.e. the packaging) or the sensitivity of the pressure sensor has to be trimmed to the application. The designer or manufacturer of chemical sensor on the other hand, has to cope not only with different applications, a pH sensor, for example, but also with the variety of chemicals that have to be measured.

In order to be able to sell the large quantities of sensors that have to be put on the market and to benefit from the low price of planar sensors, the technological base for all different types of chemical sensor has to be identical. Also, the customer expects an identical signal output, connector and user interface for the different sensors. In this book, these aspects have been taken into account. Firstly, the different technologies described are universal and can be applied to a large variety of sensors. Secondly, by adapting smart sensor techniques, the output of, and the interfacing to, all types of sensors can be made identical.

So, there are several indications that microelectrochemical biosensors will soon become a commercial reality. However, the ultimate goal (or is it a dream?) of all researchers working on glucose sensors—'the realization of an in vivo glucose sensor with an operational lifetime of more than one year'—has not yet been achieved. This requires intensive research, not only on planar production techniques, but also on improved biochemical systems, biocompatibility schemes, interfacing circuits and packaging techniques. There is still a long way to go, but the first steps have certainly been taken.

LIST OF REFERENCES

Acheson 1991 *Product Data Sheets on screen-printable Ag, Ag/AgCl and C pastes* (Acheson Colloiden B V, Scheemda, The Netherlands)

Adams R N 1969 Electrochemistry at solid electrodes *Monographs in electroanalytical chemistry and electrochemistry* ed A J Bard (New York: Marcel Dekker)

Aizawa M 1983 Molecular recognition and chemical amplification of biosensors *Proc. 1st Int. Meeting on Chemical Sensors (Fukuoka, 1983)* pp 683–92

Albisser A M 1979 Devices for the control of diabetes mellitus *Proc. IEEE* **67** 1308–19

Amine A, Kauffmann J-M and Patriarche G J 1991 Amperometric biosensors for glucose based on carbon paste modified electrodes *Talanta* **38** 107–10

Aoki K, Tokuda K and Matsuda H 1987a Theory of chronoamperometric curves at microband electrodes *J. Electroanal. Chem.* **225** 19–32

—— 1987b Derivation of an approximate equation for chronoamperometric curves at microband electrodes and its experimental verification *J. Electroanal. Chem.* **230** 61–7

Atkinson J K, Shahi S S, Varney M and Hill N 1991 A thick film electrochemical instrument *Sensors and Actuators* B **4** 175–81

Attridge J W, Daniels P B, Deacon J K, Robison G A and Davidson G P 1991 Sensivity enhancement of optical immunosensors by the use of surface plasmon resonance fluoroimmunoassay *Biosensors and Bioelectronics* **6** 201–14

Bard A J and Faulkner L R 1980 *Electrochemical methods: fundamentals and applications* (New York: Wiley)

Bartlett P N and Whitaker R G 1987 Electrochemical immobilisation of enzymes *J. Electroanal. Chem.* **224** 27–48

Bean K E 1978 Anisotropic etching in silicon *IEEE Trans. Electron Devices* **ED-25** 1185–93

Belford R E, Owen A E and Kelly R G 1987 Thick-film hybrid pH sensors *Sensors and Actuators* **11** 387–98

282   *List of References*

Berg P and Näbauer T 1988 Encapsulation of ISFETs *EEC Workshop— The interface between biology and sensors (Milton Keynes, UK, 1988)* (Cranfield Biotechnology Centre)

Bergveld P 1970 Development of an ion-sensitive solid-state device for neurophysiological measurements *IEEE Trans. Biomed. Devices* BME-17 70–1

—— 1972 Development, operation and application of the ion-sensitive field-effect transistor as a tool for electrophysiology *IEEE Trans. Biomed. Devices* BME-19 342–51

—— 1986 Chemically sensitive electronic devices: the interface between microanalysis and microelectronics *Proc. 2nd Int. Meeting on Chemical Sensors (Bordeaux, 1986)* pp 49–58

—— 1991 Future applications of ISFETS *Sensors and Actuators* B 4 125–33

Bio-Metric Systems Inc 1991 *PhotoLink, surface modification and immobilisation technology* Form No. PL0001-01 (Eden Prairie, MN: Bio–Metric Systems Inc) pp 1191

Blackburn G and Janata J 1982 The suspended mesh ion selective field effect transistor *J. Electrochem. Soc.* 129 2580–7

Blennemann H, Bousse L, Bowman L and Meindl J D 1987 Silicon chemical sensors with microencapsulation of ion-selective membranes *IEEE Proc. Transducers'87 (Tokyo)* pp 120–5

Bockris J O M and Reddy A K N 1977 *Modern Electrochemistry* vols 1 and 2 (New York: Plenum/Rosetta)

Bousse L J, Bergveld P and Geeraedts H J M 1986 Properties of Ag/AgCl electrodes fabricated with IC-compatible techniques *Sensors and Actuators* 9 179–97

Bowman L and Meindl J D 1986 The packaging of implantable integrated sensors *IEEE Trans. Biomed. Devices* BME-33 248–55

Boyle W and Smith G 1970 Charge-coupled semiconductor devices *Bell Systems Tech. J.* 49 587–93

Brinkman E 1989 Application of hydrogels for medical sensor catheters *PhD Thesis* Twente University

Britz D 1978 iR elimination in electrochemical cells *J. Electroanal. Chem.* 88 309–52

Brooks A D and Donovan R P 1972 Low-temperature electrostatic silicon-to-silicon seals using sputtered borosilicate glass *J. Electrochem. Soc.* 119 545–6

Buck R P and Hackleman D E 1977 Field effect potentiometric sensors *Anal. Chem.* 49 2315–21

Buckles S L 1987 Use of argon plasma for cleaning hybrid circuits prior to wire bonding *ISHM Proc. EMC'87 (Bournemouth, UK)* pp 78–81

Burke L D 1980 *Electrodes of conductive metallic oxides* ed S Trassatti (New York: Elsevier)

Butler J and Cobbold R 1974 A multicathode oxygen sensor fabricated using integrated circuit techniques *Digest of papers of the 5th CMBEC (Montreal, 1974)*

Buxbaum E *et al* 1989 New microminiaturised glucose sensor using covalent immobilisation techniques *Proc. Tranducers'89 (Montreux)*

Callewaert L and Sansen W 1984 Process listing 3 μm 3poly CMOS *Internal report K U Leuven* ESAT

Cardosi M and Turner A 1987 Glucose sensors for the management of diabetes mellitus *The Diabetes Annual 3* ed K Alberti and L Krall (New York: Elsevier) pp 560–77

Cass *et al* 1984 Ferrocene-mediated enzyme electrode for amperometric determination of glucose *Anal. Chem.* **56** 668–71

Cha C S, Shao M J and Liu C C 1990a Problems associated with the miniaturization of a voltammetric oxygen sensor: chemical crosstalk among electrodes *Sensors and Actuators* B **2** 239–42

—— 1990b Electrochemical behaviour of microfabricated thick-film electrodes *Sensors and Actuators* B **2** 277–81

Chan L T, Foley D L, Yao S J, Krupper M A and Wolfson S K 1987 A method for electrochemical glucose sensing employing electrochemical regeneration by voltage pulsing *Proc. EMBS'87 (Boston, MA)* pp 794–5

Christian G D and O'Reilly J E 1986 *Instrumental analysis* (Boston, MA: Allyn and Bacon)

Churchouse S J, Battersby C M, Mullen W H and Vadgama P M 1986 Needle enzyme electrodes for biological studies *Biosensors* **2** 325–42

Claes A and Maene N 1987 Realisatie van voltammetrische dikke film sensoren *MS thesis* K U Leuven ESAT–MICAS

Clark L C 1956 Monitor and control of blood and tissue oxygen tension *Trans. Am. Soc. Artif. Int. Organs* **2** 41–8

Clark L C and Duggan C A 1982 Implanted electroenzymatic glucose sensors *Diabetes Care* **5** 174–80

Cline H E and Anthony T R 1976 High-speed droplet migration in silicon *J. App. Phys.* **47** 2325–31

Clinica 1990 MediSense to launch Satellite system *Clinica* **402** 17

Coleman M V and Winster A E 1981 Silver migration in thick film conductors and chip attachment resins *Microelectronics J.* **12**

De Wachter D 1990 Smartie, een slimme voltammetrische sensorinterface *MS thesis* K U Leuven ESAT–MICAS

Deacon J K *et al* 1991 An assay for human chorionic gonadotrophin using the capillary fill immunosensor *Biosensors & Bioelectronics* **6** 193–9

Degani Y and Heller A 1987 Direct electrical communication between chemically modified enzymes and metal electrodes *J. Phys. Chem.* **91** 1285–9

Devanathan D and Carr R 1980 Polymeric conformal coatings for implantable electronic devices *IEEE Trans. Biomed. Devices* **BME-27** 671–4

Dicks J M, Hattori S, Karube I, Turner A P F and Yokozawa T 1989

Ferrocene modified polypyrrole with immobilised glucose oxidase and its application in amperometric glucose microsensors *Ann. Biol. Clin.* 47 607–19

Dionex 1989 Analysis of carbohydrates by anion exchange chromatography with pulsed amperometric detection *Techical Note* TN 20 (Sunnyvale, CA: Dionex Corporation)

Donaldson P 1976 The encapsulation of microelectronic devices for long term surgical implantation *IEEE Trans. Biomed. Devices* **BME-23** 281–5

—— 1991 Aspects of silicone rubber as an encapsulant for neurological prostheses. Part 1: osmosis *Med. Biol. Eng. Comp.* 29 34–9

Dumschat C *et al* 1990 Encapsulation of chemically sensitive field-effect transistors with photocurable epoxy resins *Sensors and Actuators* B 2 271–6

Dupont 1989 TEFLON AF, a new generation of Teflon fluorocarbon resins for high performance *Technical information from Dupont* (Wilmington, DE: Speciality Business Centre)

Eddowes M J 1987 Response of an enzyme-modified pH-sensitive ion-selective device *Sensors and Actuators* 11 265–74

Eden G, Inbar G, Timor-Tritsch I and Bicher H 1975 Miniaturised electrode for on-line pO_2 measurements *IEEE Trans. Biomed. Devices* **BME-22** 275–80

EG&G PARC 1982 Basics of voltammetry and polarography *Application note* P-2, and also P-3, D-2, F-2, S-7 and AC-1 (EG&G Princeton Applied Research)

Enfors S O 1981a Oxygen-stabilised enzyme electrode for D-glucose analysis in fermentation broths *Enzyme Microb. Technol.* 3 29–32

—— 1981b Enzyme electrode using electrolytic oxygen *European patent specification* 35480

Engels J M, Kimmich H P, Kuypers M H, Maas H G and Vervoort M P 1982 A disposable oxygen sensor for biomedical applications *Solid-state sensors and transducers, Summer Course 1982* ed W Sansen and J Van der Spiegel (K U Leuven ESAT)

Engels J M and Kuypers M H 1983 Medical applications of silicon sensors *J. Phys. E: Sci. Instrum.* 16 987–94

Erlich D J, Silversmith D J, Mountain R W and Tsao J 1982 Fabrication of through-wafer via conductors in Si by laser photochemical processing *IEEE Trans. Comp. Hybr. Man. Tech.* **CHMT-5** 520–1

Evans D H, O Connell K M, Petersen R A and Kelly M J 1983 Cyclic voltammetry *J. Chem. Ed.* 60 290–3

Ewing G W 1975 *Instrumental methods of chemical analysis* (Kogakusha, Tokyo: McGraw–Hill)

Fog A and Buck R P 1984 Electronic semiconducting oxides as pH-sensors *Sensors and Actuators* 5 137–45

Frary J M and Seese P 1981 Lift-off techniques for fine line metal patterning *Semiconductor International* pp 72–88

Freiser H 1978 *Ion-selective electrodes in analytical chemistry 1* (New York: Plenum)

——— 1980 *Ion-selective electrodes in analytical chemistry 2* (New York: Plenum)

Garcia O, Celen J, Sansen W and Colin F 1983 Electrode enzymatic implantable *Journées d'Electrochémie* 8–10

Gernet S, Koudelka M and de Rooij N F 1989a A planar glucose enzyme electrode *Sensors and Actuators* 17 537–40

——— 1989b Fabrication and characterization of a planar electrochemical cell and its application as a glucose sensor *Sensors and Actuators* 18 59–70

Glembocki O J and Stahlbush R E 1985 Bias-dependent etching of silicon in aqueous KOH *J. Electrochem. Soc.* 132 145–51

Gough D A, Leypolt J K and Armour J C 1982 Progress toward a potentially implantable enzyme-based glucose sensor *Diabetes Care* 5 190–8

Grebe K, Ames I and Ginzberg A 1974 Masking of deposited thin films by means of an aluminium-photoresist composite *J. Vac. Sci. Technol.* 11 458–60

Grimm L, Hilke K J and Scharrer E 1983 The mechanism of the cross linking of poly(vinyl alcohol) by ammonium dichromate with UV light *J. Electrochem. Soc.* 130 1767–71

Grove A S 1967 *Physics and technology of semiconductor devices* (New York: Wiley)

Guckel H, Burns D W, Rutigliano C R, Showers D K and Uglow J 1987 Fine grained polysilicon and its application to planar pressure transducers *IEEE Proc. Transducers'87 (Tokyo)* pp 277–82

Guilbault G, Kauffmann J-M and Patriarche G 1991 Immobilsed enzyme electrodes as biosensors *Protein immobilisation: fundamentals and applications* ed R F Taylor (New York: Dekker) pp 209–36

Gumbrecht W *et al* 1991 Monitoring of blood pO2 with a thin film amperometric sensor *Technical Digest Transducers'91 (San Fransisco)* pp 85–7

Hanazato Y, Nakako M, Maeda M and Shiono S 1986 Application of water-soluble photocrosslinkable polymer to enzyme membrane for FET-biosensor *Proc. 2nd Int. Meeting on Chemical Sensors (Bordeaux, 1986)* pp 576–9

Hardy M and Townsend R 1988 Separation of positional isomers of oligosaccharides and glycopeptides by high-performance anion-exchange chromatography with pulsed amperometric detection *Proc. Natl Acad. Sci. USA* 85 3289–93

Harmer M A and Hill A O 1985 The direct electrochemistry of redox proteins at metal oxide electrodes *J. Electroanal. Chem.* 189 229–46

Hatzakis M, Canavello B J and Shaw J M 1980 Single step optical lift-off process *IBM J. Res. Develop.* (July 1980)

Haviland J, Krahn D and Johnson W 1991 Semi-custom integrated silicon sensor development array *Technical Digest Transducers'91 (San Fransisco)* pp 436–9

Heintz F and Zabler E 1987 Motor vehicle sensors based on film-technology *SAE Proc. Sensors and Actuators'87 (Detroit)* **SP693** 115–24

Henning W 1991 Bus systems *Sensors and Actuators A* **25–27** 109–13

Higgins *et al* 1985 Sensor for components of a liquid mixture *US patent* 4 545 382

Hirata M, Suwazono S and Tanigawa H 1985 A silicon diaphragm formation for pressure sensor by anodic oxidation etch-stop *IEEE Proc. Transducers'85 (Philadelphia)* pp 287–90

—— 1987 Diaphragm thickness control in silicon pressure sensors using an anodic oxidation etch-stop *J. Electrochem. Soc.* **134** 2037–41

Ho N J, Kratochvil J, Blackburn G F and Janata J 1983 Encapsulation of polymeric membrane-based ion-selective field effect transistors *Sensors and Actuators* **4** 413–21

Hoare P 1968 *The electrochemistry of oxygen* (New York: Wiley)

Honig E P 1980 A miniature oxygen sensor *Proc. NNV-symposium, Vaste stof sensoren* (Deventer: Kluwer) pp 141–4

—— 1990 Private communication (Philips, Eindhoven)

Hosticka B, Fichtel J and Zimmer G 1984 Integrated monolithic temperature sensors for acquisition and regulation *Sensors and Actuators* **6** 191–200

IC Sensors 1987 *IC Sensors technical documentation* (Milpitas, CA: IC Sensors)

Ichimura K and Watanabe S 1982 Preparation and characteristics of photocrosslinkable polyvinyl alcohol *J. Polymer Sci.* **20** 1419–23

Jackson T N, Tischler M A and Wise K D 1981 An electrochemical p-n junction etch-stop for the formation of silicon microstructures *IEEE Electron Devices Letters* **EDL-2** 44–5

Jacobs P 1989 Continue rendementsbepaling van een kunstnierbehandeling in vivo *MS Thesis* K U Leuven ESAT–MICAS

Jacobs P, Suls J and Sansen W 1990 A planar conductometric sensor showing excellent polarisation impedance characteristics *IEEE Proc. EMBS* **12** 1484–5

Jaeger R C 1988 *Introduction to microelectronic fabrication* (Reading, MA: Addison–Wesley)

Janata J 1986 Chemical selectivity of field effect transistors *Proc. 2nd Int. Meeting on Chemical Sensors (Bordeaux, 1986)* pp 25–31

—— 1990 Chemical sensors *Anal. Chem.* **62** 33R–44R

Janata 1992 Chemical Sensors *Anal. Chem.* **64** 196R–219R

Janata J and Bezegh A 1988 Chemical sensors *Anal. Chem.* **60** 62R–74R

Janata J J and Huber R J 1980 *Chemically sensitive field-effect transistors* in Freiser 1980

Johnson D 1986 Carbohydrate detection gains potential *Nature* **321** 451–2

Josowicz M, Janata J and Levy M 1988 Electrochemical pretreatment of thin film platinum electrodes *J. Electrochem. Soc.* **135** 112–5

K U Leuven–ESAT 1983 Process listing of the double poly p-well LUVC-MOS3 process *K U Leuven–ESAT Internal report*

Karagounis V, Lun L and Liu C 1986 A thick film multiple component cathode three-electrode oxygen sensor *IEEE Trans. Biomed. Devices* **BME-33** 108–12

Karube I and Suzuki H 1988 Miniaturised oxygen electrode and miniaturised biosensor and production process thereof *European patent application* 284 518

Karube I and Tamiya E 1986 Micro-biosensor for clinical analysis *Proc. 2nd Int. Meeting on Chemical Sensors (Bordeaux, 1986)* pp 588–91

Kelly R G 1977 Microelectronic approaches to solid-state ion-selective electrodes *Electrochim. Acta* **22** 1–8

Kendall D L 1975 On etching very narrow grooves in silicon *Appl. Phys. Lett.* **26** 195–8

Kern W 1978 Chemical etching of silicon, germanium, gallium arsenide and gallium phosphide *RCA review* **39** 278–308

Kern W and Rosier R S 1977 Advances in deposition processes for passivation films *J. Vac. Sci. Technol.* **14** 1082–99

Kim S J, Kim M and Heetderks W J 1985 Laser-induced fabrication of a transsubstrate microelectrode array and its neurophysiological performance *IEEE Trans. Biomed. Devices* **BME-32** 497–502

Kimmich H P and Kreuzer F 1969 *Oxygen pressure recordings in gases, fluids and tissues* (Basle: Karger) pp 100–10

Kimura J, Kuriyama T and Kawana Y 1985 An integrated SOS/FET multi-biosensor and its application to medical use *Proc. Transducers'85 (Philadelphia)* pp 152–5

Kimura J, Murikami T, Kuriyama T and Karube I 1988 An integrated multibiosensor for simultaneous amperometric and potentiometric measurement *Sensors and Actuators* **15** 435–43

Kloeck B 1989 Design, fabrication and characterisation of piezoresistive pressure sensors, including the study of electrochemical etch-stop *PhD Dissertation* University of Neuchatel

Kloeck B, Collins S, de Rooij N F and Smith R 1989 Study of electrochemical etch-stop for high-precision thickness control of silicon membranes *IEEE Trans. Electron Devices* **ED-36** 663–9

Kloeck B and de Rooij N F 1987 A novel four electrode electrochemical etch-stop method for silicon membrane formation *IEEE Proc. Transducers'87 (Tokyo)* pp 116–9

Knecht T A 1987 Bonding techniques for solid state pressure sensors *IEEE*

Proc. Transducers'87 (Tokyo) pp 95–8

Kobayashi T 1985 Solid-state sensors and their applications in consumer electronics and home appliances *IEEE Proc. Transducers'85 (Philadelphia)* pp 8–12

Koopal C, de Ruiter B and Nolte R 1991 Amperometric biosensor based on direct communication between glucose oxidase an a conducting polymer inside the pores of a filtration membrane *J. Chem. Soc.* accepted for publication

Koudelka M 1986 Performance characteristics of a planar Clark-type oxygen sensor *Sensors and Actuators* 9 249–58

Koudelka M, Gernet S and de Rooij N F 1987 Voltammetry—a powerful tool for evaluation and process control of thin film electrodes *IEEE Proc. Transducers'87 (Tokyo)* pp 41–4

Koudelka M, Gernet S and de Rooij N F 1988 A planar glucose enzyme electrode *Technical Digest (Eurosensors II, Twente, The Netherlands)*

—— 1989 Planar amperometric enzyme-based glucose microelectrode *Sensors and Actuators* 18 157–65

Koudelka M and Grisel A 1985 Miniaturized Clark-type oxygen sensor *IEEE Proc. Transducers'85 (Philadelphia)* pp 418–21

Koudelka M, Rohner-Jeanrenaid F, Terrettaz J, Bobbioni-Harsch E, de Rooij N F and Jeanrenaud B 1991 In vivo behaviour of hypodermically implanted microfabricated glucose sensor *Biosensors and Bioelectronics* 6 31–6

Kulys J and D Costa E J 1991 Printed amperometric sensor based on TCNQ and cholinesterase *Biosensors and Bioelectronics* 6 109–15

Kuriyama T, Kimura J and Kawana Y 1985 A single chip biosensor *NEC Research and Development* 78 1–5

Kuriyama T, Nakamoto S, Kawana Y and Kimura J 1986 New fabrication methods of enzyme immobilized membrane for ENFET *Proc. 2nd Int. Meeting on Chemical Sensors (Bordeaux, 1986)* pp 568–71

Kuypers M H 1985 Clark cell with hydropylic polymer layer *US patent* 4 492 622

—— 1988 A polarographic-amperometric three-electrode sensor *European patent application* 88 112 641.1

—— 1989 *Private communication* (PPG Hellige, Best, NL)

—— 1990 *Private communication on the development of oxygen sensors* (Dräger Medical Electronics, Best, NL)

Kuypers M H and Steeghs G 1989 Process and sensor for measuring the glucose content of glucose-containing fluids *European patent application* 89 108 264.6

Kuypers M H, Steeghs G and Brinkman E 1988 Method of providing a substrate with a layer comprising a polyvinyl based hydrogel and a biochemically active material *European patent application* 88 116 789.4

Lambrechts M 1985 Realizatie van geintegreerde, planaire voltammetrische sensoren met behulp van micro-elektronica en dikke film technieken *Annual report for the IWONL fellowship* (K U Leuven–ESAT)

—— 1989a Planar voltammetric sensors *PhD thesis* K U Leuven ESAT-MICAS

—— 1989b On the use of impression material for sensor applications *Internal Report K U Leuven ESAT-MICAS*

Lambrechts M, Suls J and Sansen W 1986a Silver multi-layers for planar Ag/AgCl reference electrodes *IEEE Proc. Solid-State Sensors Workshop (Hilton Head SC, 1986)*

—— 1986b Corrosion resistance of silver thin film layers for planar voltammetric sensors *Proc. 2nd Int. Meeting on Chemical Sensors (Bordeaux, 1986)* pp 572–5

—— 1987 A thick film glucose sensor *Proc. EMBS'87 (Boston)* pp 798–9

—— 1988 Thick film voltammetric sensors based on ruthenium dioxide *Sensors and Actuators* 13 287–92

Lauks I 1981 Polarizable electrodes, part II *Sensors and Actuators* 1 393–402

Lauks I, Van der Spiegel J, Sansen W and Steyaert M 1985 Multispecies integrated electrochemical sensors with on-chip CMOS circuitry *IEEE Proc. Transducers'85 (Philadelphia)* pp 122–4

Lewandowski J, Malchesky P, Moorman M, Nalecz M and Nose Y 1986 Electrocatalytic determination of glucose in a Ringer's solution by pulse-voltammetry *IEEE Trans. Biomed. Devices* BME-33 147–52

Lewandowski J J, Malchesky P S, Zborowski M and Nose Y 1987 Assessment of microelectronic technology for fabrication of electrocatalytical glucose sensor *Proc. EMBS'87 (Boston)* pp 784–5

Lewis E 1985 Biological sensors *IEEE Proc. Transducers'85 (Philadelphia)* pp 13–6

Ligtenberg H 1987 Baseline drift mechanism of SiO_2/Al_2O_3 pH ISFETs and some improvements by pulsed excitation *IEEE Proc. Transducers'87 (Tokyo)* pp 747–50

Loeb W E 1977 Parylene removal with oxygen plasmas *Union Carbide Corporation Report* no 18 pp 1–5

Lowe C R 1985 An introduction to the concepts and technology of biosensors *Biosensors* 1 3–16

Lundström I 1981 Hydrogen sensitive MOS-structures. Part 1: principles and applications *Sensors and Actuators* 1 403–26

Lundström I and Södderberg D 1982 Hydrogen sensitive MOS-structures. Part 2: characterisation *Sensors and Actuators* 2 105–38

Mallinckrodt 1990 *GEM Systems, Technical Summary* Rev. 3/90 (Ann Arbor, MI: Mallinckrodt Sensors Systems)

Maloy J T 1991 Factors affecting the shape of current–potential curves *J. Chem. Ed.* 60 285–9

Maszara W P 1991 Silicon-on-insulator by wafer bonding: a review *J. Elec-*

trochem. Soc. **138** 341–7

Matsuo T and Wise K D 1974 An integrated field-effect electrode for biopotential recording *IEEE Trans. Biomed. Devices* **BME-21** 485–7

Matthews D R, Holman R R, Bown E, Steemson J, Watson A, Hughes S and Scott D 1987 Pen-sized digital 30-second blood glucose meter *Lancet* pp 778–9

McKean B D and Cough D A 1988 A telemetry-instrumentation system for chronically implanted glucose and oxygen sensors *IEEE Trans. Biomed. Devices* **BME-35** 526–32

McKinnon G, Tieszen K, McMurray J and Dornan T 1991 An assesment of the satellite G electrochemical glucose sensor using venous and capillary blood *Poster presented at the 14th Int. Diabetes Federation Congress (Washington DC, 1991)*

Mead C and Conway L 1986 *Introduction to VLSI Systems* (London: Addison–Wesley)

MediSense 1991 *Made to measure. A complete series of blood glucose measuring systems from MediSense* (Nieuwegein, The Netherlands: MediSense Europe B V)

Microsensors Inc. 1988 *Catalogue and general information. Chemical microsensors and automatic vapour generation systems* (Fairfax, VA: Microsystems Inc)

Middelhoek S and Audet S A 1987 Silicon sensors: full of promises and pitfalls *J. Phys. E: Sci. Instrum.* **20** 1080–6

Middelhoek S and Noorlag D 1981 Signal conversion in solid-state transducers *Proc. Symposium Solid State Transducers (Boston, 1981)*

Miyahara Y, Matsu F and Moriizumi T 1983 Micro enzyme sensors using semiconductor and enzyme immobilization techniques *Proc. 1st Int. Meeting on Chemical Sensors (Fukuoka, 1983)* pp 502–6

Miyahara Y and Moriizumi T 1985 Integrated enzyme FET's for simultaneous detections of urea and glucose *Sensors and Actuators* **7** 1–10

Morff *et al* 1990 Microfabrication of reproducible, economical electroenzymatic glucose sensors *IEEE Proc. EMBS* **12** 483–4

Moriizumi T and Miyahara Y 1985 Monolithic multi-function ENFET biosensors *Proc. Transducers'85 (Philadelphia)* pp 148–51

Moriizumi T, Takatsu I and Ono K 1986 Solid state biosensors using thin-film electrodes *Proc. 2nd Int. Meeting on Chemical Sensors (Bordeaux, 1986)* pp 647–50

Moss S D , Janata J and Johnson C C 1975 Potassium Ion-Selective Field Effect Transistor *Anal. Chem.* **47** 2238–43

Muller R S 1989 Microdynamics *Proc. Tranducers'89 (Montreux)* pp 25–30

Murikami T, Kimura J and Kuriyama T 1987 An integrated multi-biosensor for simultaneous amperometric and potentiometric measurements *IEEE Proc. Transducers'87 (Tokyo)* pp 804–7

Näbauer A, Berg P, Ruge I and Riedlberger F 1986 Ein Transducer für Biosensoren auf der Basis eines Feldeffekttransistors *Sensoren-Technologie und Anwendung, NTG-Fachberichte 93* (Bad Nauheim: VDE-Verlag Gmbh) pp 39–46

NovaSensor 1987 *Technical documentation* (Fremont, CA: NovaSensor)

Obermeier E 1985 Polysilicon layers lead to a new generation of pressure sensors *IEEE Proc. Transducers'85 (Philadelphia)* pp 430–3

Oesch U and Simon W 1987 Opportunities of planar ISE membrane technology *IEEE Proc. Transducers'87 (Tokyo)* pp 755–9

Oldham K B 1981 Edge effects in semiinfinite diffusion *J. Electroanal. Chem* **122** 1–17

Olechno J, Carter S, Edwards W D and Gillen 1987 Developments of chromatograpic determination of carbohydrates *Am. Biotechnol. Lab.* (September/October 1987)

Osteryoung J 1983 Pulse voltammetry *J. Chem. Ed.* **60** 296–8

Otagawa T and Madou M 1989 Surface type microelectronic gas and vapor sensor *European patent application* 0 288 780

Pace S J 1981 Surface modification and commercial application *Sensors and Actuators* **1** 475–527

——— 1986 *Private communication*

Pace S J and Jensen M A 1986 Thick-film multi-layer pH sensor for biomedical applications *Proc. 2nd Int. Meeting on Chemical Sensors (Bordeaux, 1986)* pp 557–61

Pace S J, Zarzycki P P, McKeever R T and Pelosi L 1985 A thick-film multi-layered oxygen sensor *Proc. Transducers'85 (Philadelphia)* pp 406–9

Palik E D, Bermudez V M and Glembocki O J 1985 Ellipsometric study of the etch-stop mechanism in heavily doped silicon *J. Electrochem. Soc.* **131** 135–41

Palik E D, Glembocki O J and Heard I 1987 Study of bias-dependent etching of Si in aqueous KOH *J. Electrochem. Soc.* **134** 404–9

Parker D 1987 Sensors for monitoring blood gases in intensive care *J. Phys. E: Sci. Instrum.* **20** 1103–12

Paszczynski S, Lambrechts M, Sansen W and Suls J 1986 Neue Moglichkeiten der Anwendung der Dickschichtentechnologie auf dem Gebiet der Sensortechnik *Proc. Hybrid Microelectronics Conf. (Gera, Germany, 1986)*

Petersen K E 1982 Silicon as a mechanical material *Proc. IEEE* **70** 420–57

Petersen K E *et al* 1991 Surface micromachined structures fabricated with silicon fusion bonding *Technical Digest Transducers'91 (San Fransisco)* pp 397–9

Peterson W M and Wong R V 1981 Fundamentals of stripping voltammetry *Am. Lab.* (November 1981)

Peura R and Mendelson Y 1984 Blood glucose sensors: an overview *Proc. IEEE/NSF Symposium on Biosensors*

Pfleiderer H J, Wieder A W and Hart K 1989 BICMOS for high performance analog and digital circuits *Proc. ESSCIRC'89 (Vienna)*

Pharmacia 1991 Brochure on BIAcore system *Application note* 101, 201, 202 (Pharmacia Biosensor A B, Uppsala, Sweden)

Pickup J and Rothwell D 1984 Technology and the diabetic patient *Med. and Biol. Eng. and Comput.* **22** 385–400

Pickup J, Shaw G and Claremont D 1988 Implantable glucose sensors: choosing the appropriate sensing strategy *Biosensors* **3** 335–46

Pilf C *et al* 1989 A new type of thin-film microelectrode for in vivo voltammetry *Technical Digest Transducers'89 (Montreux)* p 289

Polak A J, Beuhler A J and Petty-Weeks S 1985 Hydrogen sensors based on proton conducting polymers *IEEE Proc. Transducers'85 (Philadelphia)* pp 85–8

Prohaska O, Goiser P, Jachomowicz A, Kohl F and Olcaytug F 1986 Miniaturized Chamber-type electrochemical call for Medical Applications *Proc. 2nd Int. Meeting on Chemical Sensors (Bordeaux, 1986)* p 652

Prohaska O *et al* 1987 Multiple chamber-type probe for biomedical application *IEEE Proc. Transducers'87 (Tokyo)* pp 812–3

Prudenziati M 1987 Screen printed and fired sensors for physical and chemical quantities *IEEE Proc. Transducers'87 (Tokyo)* pp 85–90

—— 1991 Thick film technology *Sensors and Actuators* A **25–27** 227–34

Puers R and Sansen W 1989 Compensation structures for convex corner micromachining in silicon *Proc. Tranducers'89 (Montreux)* pp 1036–41

Rikoski R A 1973 *Hybrid microelectronic circuits: the thick film* (New York: Wiley)

Ripka G and Harsanyi G 1985 Electrochemical migration in thick film IC-s *Electrocomp. Sci. and Tech.* **11** 281–90

Rocklin R, Henshall A and Rubin R 1990 A multimode electrochemical detector for non-UV-absorbing molecules *American Laboratory* pp 34–48

Rolfe P 1988 Haemocompatibility of intra-vascular senors *EEC Workshop— The interface between biology and sensors (Milton Keynes, UK, 1988)* (Cranfield Biotechnology Centre)

Ross P N 1979 Structure sensitivity in the electrocatalytic properties in Pt *J. Electrochem. Soc.* **126** 67–82

Ruger P, Bilitewski U and Schmid R D 1991 Glucose and ethanol biosensors based on thick film technology *Sensors and Actuators* B **4** 267–71

Sansen W 1982 On the integration of an Internal Human Conditioning System *IEEE J. Solid-state Circuits* **SC-17** 513–22

—— 1987 Integrated low-noise amplifiers in CMOS technology *Nucl. Instrum. Methods* A **253** 427–33

Sansen W, Celen J, Colin F and Garcia O 1983 Implantable glucose measurement system *Proc. 5th Ann. Conf. IEEE EMBS (Columbus, OH, 1983)* pp 341–4

Sansen W and Lambrechts M 1985 Glucose sensor with telemetry system *Implantable sensors for closed-loop prostetic systems* ed Wen H Ko (Mount Kisco, NY: Futura) pp 167–75

Sansen W, Lambrechts M and Suls J 1985 Fabrication of voltammetric sensors with planar techniques *IEEE Proc. Transducers'85 (Philadelphia)* pp 344–7

Sansen W, Steyaert M, Lambrechts M and Vriens K 1987 Integrated signal-processing electronics for planar electrochemical sensors *Proc. ISCAS'87 (Philadelphia)* pp 225–8

Sansen W, Suls J, Lambrechts M, Claes A, Jacobs P and Kuypers M M 1982 Anaerobic operation of a glucose sensor by use of pulse techniques *Proc. Biosensors'92 (Geneva)* (Amsterdam: Elsevier) pp 156–62

Santiago J, Clemens A, Clarke W and Kipnis D Closed-loop and open-loop devices for blood glucose control in normal and diabetic subjects *Diabetes* **28** 71–84

Sarangapani *et al* 1987 Electrocatalytic glucose sensor *Proc. EMBS'87 (Boston)* 796–7

Schnable G L, Kern W and Comizzoli R B 1975 Passivation coatings on silicon devices *J. Electrochem. Soc.* **122** 1092–1102

Schnakenberg U, Benecke W and Lange P 1991 TMAHW etchants for silicon micromachining *Technical Digest Transducers'91 (San Fransisco)* pp 815–8

Schnakenberg U, Benecke W and Lochel B 1990 NH$_4$OH-based etchants for silicon micromachining *Sensors and Actuators* A **21–23** 1031–5

Schneider B, Prohaska O and Daroux M 1989 Microminiature enzyme sensor for glucose and lactate based on chamber oxygen electrodes *Technical Digest Transducers'89 (Montreux)* p 318

Schoen P E *et al* 1990 Lithographic definition of biomaterials on surfaces *IEEE Proc. EMBS* **12** p 1793

Scholze J, Hampp N and Brauchle C 1991 Enzymatic Hybrid biosensors *Sensors and Actuators* B **4** 211–5

Schwartz B and Robbins H 1976 Chemical etching of silicon (IV. Etching technology) *J. Electrochem. Soc.* **123** 1903–9

Seidel H 1987 The mechanism of anisotropic silicon etching and its relevance for micromachining *IEEE Proc. Transducers'87 (Tokyo)* pp 120–5

Seidel H and Gsepregi L 1983 Three-dimensional structuring of silicon for sensor applications *Sensors and Actuators* **4** 455–63

Seipler D G 1987 Sensors—integration in thick film hybrid circuits *ISHM Proc. EMC'87 (Bournesmouth, UK)*

Shafer M W *et al* 1979 Preparation and characterisation of RuO$_2$ crystals *J. Electrochem. Soc.* **126** 1625–8

Sharp D J 1979 Corrosion inhibition in sputter-deposited thin-film systems using an intermediary layer of palladium *J. Vac. Sci. Technol.* **16** 204–7

Sheppard N 1991 Design of a conductimetric microsensor based on reversibly swelling polymer hydrogels *Technical Digest Transducers'91 (San Fransisco)* pp 773–6

Shichiri M, Kawamori R, Goriga Y, Yamasaki Y, Hakui N and Abe H 1983 Glycaemic control in pancreatectomised dogs with a wearable endocrine pancreas *Diabetologia* **24** 179–84

Shichiri M, Kawamori R, Yamasaki Y, Hakui N and Abe H 1982 Wearable artificial endocrine pancreas with needle-type glucose sensor *Lancet* pp 1129–31

Shumilova N A and Zhutaeva G V 1987 Silver *Encyclopedia of electrochemistry of the elements* vol 8, ed A Bard (New York: Dekker)

Sibbald A 1985 A chemical sensitive integrated circuit, the operational transducer *Sensors and Actuators* **7** 23–38

Sibbald A, Covington A K and Carter R F 1985 On-line patient-monitoring system for the simultaneous analysis of blood K^+, Ca^{2+}, Na^+ and pH using quadruple-function Chemfet integrated-circuit sensor *Med. and Biol. Eng. Comp.* **23** 329–38

Siu W and Cobbold R 1976 Characteristics of a multicathode polarographic oxygen electrode *Medical and Biological Engineering* pp 109–20

Skoog D A and West D M 1976 *Fundamentals of analytical chemistry* (New York: Holt, Rinehart and Winston)

Smith R L and Collins S D 1988 Micromachined packaging for chemical microsensors *IEEE Trans. Electron Devices* **ED-35** 787–92

Smith R L, Kloeck B, De Rooij N J and Collins S D 1987 The potential dependence of silicon anisotropic etching in KOH at 60 °C *J. Electroanal. Chem* **238** 103–13

Solsky R 1988 Ion-selective electrodes *Anal. Chem.* **60** 34R–41R

Spencer W J 1981 A review of programmed insulin delivery systems *IEEE Trans. Biomed. Eng.* **BME-28** 237–51

Steyaert M 1987 Monolithic low-power data-acquisition system for biomedical purposes *PhD dissertation* K U Leuven ESAT-MICAS

Steyaert M, Lambrechts M and Sansen W 1987 Noise power spectrum density analysis of planar Ag/AgCl electrodes *Sensors and Actuators* **12** 185–92

Stryer L 1983 *Biochemistry* 2nd edn (New York: Freeman)

Suls J, Lambrechts M, Paszczynski S and Sansen W 1986 Thick film voltammetric sensors based on ruthenium dioxide *Proc. 2nd Int. Meeting on Chemical Sensors (Bordeaux)* pp 764–7

Suzuki H, Kojima N, Suguma A and Fujita S 1991 Micromachined Clark oxygen electrode *Technical Digest Transducers'91 (San Fransisco)* pp 339–42

Suzuki H, Sugama A and Kojima N 1991 Miniature Clark-type oxygen electrode with a three electrode configuration *Sensors and Actuators* B **2** 297–303

Suzuki H, Tamiya E and Karube I 1988 Fabrication of an oxygen electrode using semiconductor technology *Anal. Chem.* **60** 1078–80

Suzuki H *et al* 1989 Disposable oxygen electrode fabricated by semiconductor techniques and their application to biosensors *Technical Digest Transducers'89 (Montreux)* p 311

Sze S M 1983 *VLSI Technology* (Tokyo: McGraw–Hill)

Tabata O, Asahi R, Funabashi H and Sugiyama S 1991 Anisotropic etching of silicon in $(CH_3)_4NOH$ solutions *Technical Digest Transducers'91 (San Fransisco)* pp 811–4

Tamiya E, Karube I, Hattori S, Suzuki M and Yokoyama K 1989 Micro glucose sensors using electron mediators immobilised on a polypyrrole-modified electrode *Sensors and Actuators* **18** 297–307

Thévenot R 1988 Development of a needle-type electrochemical glucose microsensor for artificial pancreas *EEC Workshop—The interface between biology and sensors (Milton Keynes, UK, 1988)* (Cranfield Biotechnology Centre)

Thornton J 1975 Influence of substrate temperature and deposition rate on structure of thick sputtered Cu coatings *J. Vac. Sci. Technol.* **12** 830–5

Tietz N W 1986 *Textbook of clinical chemistry* (Philadelphia, PA: Saunders)

Tompsett M *et al* 1971 Charge-coupled imaging devices; experimental results *IEEE Trans. Electron. Devices* **ED-18** 992–6

Trasatti S and O Grady W E 1978 Properties and applications of RuO_2-based electrodes *Adv. Electrochem. Electrochem. Eng.* **12** 177–261

Tsakuda K, Scbata M, Maruizumi T, Miyahara Y and Miyagi H 1987 A multiple-chemfet integrated with CMOS interface circuits *IEEE Proc. Transducers'87 (Tokyo)* pp 155–8

Tseung A and Goffe R In vivo reference electrodes *Med. and Biol. Eng. and Comput.* **16** 677–80

Turner A P F 1985 Diabetes Mellitus: biosensors for research and management *Biosensors* **1** 85–115

Turner A P F, Karube I and Wilson G 1987 *Biosensors. Fundamentals and applications* (Oxford: Oxford University Press)

Turner A P F 1988 The advantages and disadvantages of mediated amperometric biosensors for in vivo monitoring *EEC Workshop—The interface between biology and sensors (Milton Keynes, UK, 1988)* (Cranfield Biotechnology Centre)

Turner A P F 1989 Current trends in biosensor research and development *Sensors and Actuators* **17** 433–50

Turner A P F 1990 Biosensors: we've had the promises—where are the products? *Int. Biotechnol. Lab. News* (September 1990) p 38

Turner R F B, Harrison D and Baltes H 1987 A CMOS potentiostat for amperometric chemical sensors *IEEE J. Solid State Circuits* **SC-23**

Turner R F B, Harrison D, Rajotte R and Baltes H 1989 Nafion polymer encapsulation of enzyme eelctrodes for glucose sensing in whole blood *Proc. Tranducers'89 (Montreux)*

UK Patent 1965 Photo-sensitive insulation based on p-xylylene polymers *UK patent* 1 141 496

1984 Project proposal GLUCOSTAT *IWONL-IRSIA proposal* Universit e Libre de Bruxelles-SAIT-K U Leuven

Updike J W and Hicks J P 1967 The enzyme electrode *Nature* **241** 986–94

Vadgama P, Tang L X and Battersby C 1988 Enzyme electrodes: modification of enzyme layer activity using membranes *EEC Workshop—The interface between biology and sensors (Milton Keynes, UK, 1988)* (Cranfield Biotechnology Centre)

Vadimski R G, Frankenthal R D and Thompson D E 1979 Ru and RuO_2 as electrical contact materials: preparation and environmental interactions *J. Electrochem. Soc.* **126** 2017–23

van den Berg A *et al* 1991 An on-wafer fabricated free-chlorine sensor *Technical Digest Transducers'91 (San Fransisco)* pp 233–6

van den Vlekkert H H 1988 A pH-ISFET and an integrated pressure sensor with back side contacts *Sensors and Actuators* **14** 165–76

van den Vlekkert H H, Arnoux C, Lomazzi P and de Rooij N F 1986 A practical approach to the application of ISFET's as accurate ion-sensors *Proc. 2nd Int. Meeting on Chemical Sensors (Bordeaux)* pp 462–6

van den Vlekkert H H, Decroux M and de Rooij N F 1987 Glass encapsulation of chemical solid state sensors based on anodic bonding *IEEE Proc. Transducers'87 (Tokyo)* pp 730–3

van der Spiegel J, Lauks I, Chan P and Babic D 1983 The extended gate chemically sensitive field effect transistor as multi-species microprobe *Sensors and Actuators* **4** 291–8

van der Zee H T, Faithfull S, Kuypers M H, Dhasmana K and Erdmann W 1985 On-line conjunctival oxygen tension as a guide to cerebral oxygenation *Anaesth. Analg.* **64** 63–7

van Hoogten L 1987 Synthese en eigenschappen van poly(hydroxyethylmethacrylaat)- membranen voor micro-electroden *Licentiaatsproefschrift* K U Leuven

van Hove N and Placke G 1988 Ontwikkeling van een CMOS compatibele planaire zuurstofsensor *MS Thesis* K U Leuven ESAT

Vossen J L and Kern W 1987 *Thin film processes* (New York: Academic)

Wang J 1986 Adsorptive stripping voltammetry *International Clinical Products Review* pp 50–9

Watson L D *et al* 1987 A microelectronic conductimetric biosensor *Biosensors* **3** 101–15

Weetall H H and Hotaling T 1975 A simple, inexpensive, disposable electrochemical sensor for clinical and immuno- assay *Biosensors* **3** 57–63

Weirauch D F Correlation of the anisotropic etching of single-crystal silicon spheres and wafers *J. Applied Physics* **46** 1478–83

Westwood W D 1974 Porosity in sputtered platinum films *J. Vac. Sci. Technol.* **11** 466–71

White R M 1985 Tactile array for robotics employing a rubbery skin and a solid-state optical sensor *IEEE Proc. Transducers'85 (Philadelphia)* pp 18–21

Wightman R M 1981 Microvoltammetric electrodes *Anal. Chem.* **53** 1125A–1134A

Woods R 1976 *Chemisorption at electrodes in Electroanalytical Chemistry* vol 9, ed A J Bard (New York: Dekker) pp1–162

Woodward J 1985 *Immobilised cells and enzymes: a practical approach* (Oxford: IRL)

Wu X P, Wu Q H and Ko W H 1986 A study on deep etching of silicon using ethylene-diamine-pyrocatechol-water *Sensors and Actuators* **9** 333–43

Yao S J, Chan L, Wolfson S K, Krupper M A and Zhou H F 1986 The low-potential approach of glucose sensing *IEEE Trans. Biomed. Devices* **BME-33** 139–46

Yao S J, Krupper M A, Wolfson S K, Li V and Wu K 1983 Low-potential voltammetry of glucose in human serum dialisate *Proc. IEEE Frontiers of Engineering and Computing in Health Care* pp 338–40

Yao S J, Li V, Dokko Y, Krupper M A and Wolfson S K 1984 Electrochemical detection of glucose at low potentials *Proc. IEEE/NSF Symposium on Biosensors* pp 75–7

Yao S J, Wolfson S K, Krupper M A and Dokko Y 1982 A potentiodynamic glucose sensor *Proc. World Congress on Medical Physics and Biomedical Engineering (Hamburg, 1982)*

Zemel J N 1975 Ion-Sensitive Field Effect Transistors and Related Devices *Anal. Chem.* **47** 255A–268A

Zhang Y *et al* 1987 Polyvinyl alcohol as a matrix for indicators for fibre optic chemical sensors *Proc. EMBS'87 (Boston)* pp 813–6

APPENDIX

Table A.1   List of Symbols

Symbol	Meaning	Typical dimension
A	electrode area	cm^2
a_j	activity of substance j	M
C	capacitance	F
C_d	double-layer capacitance	μF, $\mu F\ cm^{-2}$
C_j	concentration of species j	M, mol cm^{-3}
C^{j*}	bulk concentration of species j	M, mol cm^{-3}
$C_j(x,t)$	concentration of j at distance x at time t	M, mol cm^{-3}
C_{ox}	capacitance of the gate insulator	pF cm^{-2}
D_j	diffusion coefficient of species j	$cm^2\ s^{-1}$
d	parallel plate separation	cm
d_m	thickness of a membrane	mm
E	potential at an electrode	V
E^0	standard EMF of a half-reaction	V
	standard potential of an electrode	V
$E^{0'}$	formal potential of an electrode	V
E_t	threshold potential	V
$E_{1/2}$	half-wave potential	V
e	quantity of charge of an electron	C
F	the faraday	C
	charge on one mole of electrons	
f	(1) F/RT	V^{-1}
	(2) frequency	Hz
G	Gibbs free energy	kJ
ΔG	Gibbs free energy change in a process	kJ
ΔG^0	standard Gibbs free energy change	kJ
i, I	current	A, μA
i_l	limiting current	A, μA
$J_j(x,t)$	flux of species j at location x at time t	mol $cm^{-2}\ s^{-1}$
j	current density	A cm^{-2}
k	(1) reaction rate constant	
	(2) Boltzmann constant	J/K
L	channel length	μm

Table A.1   (continued)

mM	10^{-3} molar	
m_0	mass transfer coefficient	m s^{-1}
n	electrons per molecule oxidized or reduced	none
O	oxidized form of standard system O + $ne \to$ R	
(O)	activity of the species O	M, mol cm^{-3}
[O]	concentration of the species O	M, mol cm^{-3}
p_m	oxygen permeability of a membrane	M mm^{-1} Pa^{-1} s^{-1}
pO_2	partial oxygen pressure	Pa, mmHg
R	(1) gas constant	J mol^{-1} K^{-1}
	(2) resistance	Ω
	(3) reduced form of standard system O + $ne \to$ R	
R_{ct}	charge transfer resistance	Ω
R_Ω	ohmic solution resistance	Ω
r	radius of spherical or cylindrical electrode	cm
S	linear potential scan rate	mV s^{-1}
T	absolute temperature	K
t	time	s
t_i	ionic transfer number	none
V	volume	cm^3
V_G	gate voltage	V
V_T^*	threshold voltage of an ISFET	V
V_D	drain voltage	V
v	(1) reaction rate	mol s^{-1} cm^{-3}
	(2) hydrodynamic velocity	cm s^{-1}
W	channel width	μm
x	distance from an electrode	cm
x_1	distance of the IHP from an electrode	nm
x_2	distance of the OHP from an electrode	nm
Z	complex impedance	Ω
Z_W	Warburg impedance	Ω
z	charge on an ion in signed units	none

Table A.2

Greek symbol	Meaning	Typical dimension
α	(1) scan rate dependence	none
	(2) transfer coefficient	none
ε_0	permitivity of free space	C V^{-1} m^{-1}
ε_r	relative dielectric constant	none
ϕ	electrostatic potential	V
Γ_j	surface excess of species j	mol cm^{-3}
η	overpotential	V
μ_n	electron mobility	cm^2 V^{-1} s^{-1}
ω	angular frequency	s^{-1}
θ	phase-angle	degree, rad

Table A.3 List of abbreviations

Abbreviation	Meaning
A/D	analogue/digital
AC	alternating current
AE	auxiliary electrode
Ag/AgCl	silver/silver chloride
AWV	arbitrary wave voltammetry
BHF	buffered hydrogen fluoride
BOD	biological oxygen demand
BSA	bovine serum albumin
CA	cellulose acetate
CAD	computer aided design
CCD	charge-coupled device
CE	counter electrode
CMOS	complementary metal oxide semiconductor
CRT	cathode ray tube
CV	cyclic voltammetry
CVD	chemical vapour deposition
D/A	digital/analogue
DC	direct current
DIL	dual in line
DF	dark field
DNA	desoxyribonucleic acid
DME	dropping-mercury electrode
DPV	differential pulse voltammetry
EDP	ethylenediamine pyrocatechol
EGFET	extended gate FET
EPROM	erasable programmable read-only memory
ESCAPE	electrode system for conductimetric and potentiometric measurements
FAD	flavine adenine dinucleotide
FDA	food and drug administration (US)
FET	field effect transistor
GOD	glucose oxidase enzyme
HF	high frequency
hCG	human chorionic gonadotrophin
HMDS	hexamethyldisilazane
I/O	input/output
I/V	current/voltage
IC	integrated circuit
IHCS	internal human conditioning system
IHP	inner Helmholtz plane
IPA	2-propanol
ISE	ion-selective electrode
ISFET	ion-sensitive FET
KOH	potassium hydroxide

Table A.3   (continued)

LCD	liquid crystal display
LED	light emitting diode
LF	light field
LPCVD	low-pressure chemical vapour deposition
LSV	linear-sweep voltammetry
LTCVD	low-temperature chemical vapour deposition
MOS	metal oxide semiconductor
NHE	normal hydrogen electrode
NMOS	n-channel metal oxide semiconductor
OHP	outer Helmholtz plane
PBS	phosphate buffered saline
PCB	printed circuit board
PCM	proton conductor membrane
PECVD	plasma enhanced chemical vapour deposition
PHEMA	polyhydroxyethylmethacrylate
PMOS	p-channel metal oxide semiconductor
PSG	phosphosilicate glass
PUR	polyurethane
PVA	polyvinylalcohol
PVC	polyvinylchloride
PVP	polyvinylpyrrolidone
RE	reference electrode
RF	radio frequency
RPM	revolutions per minute
RTV	room-temperature vulcanize
SAW	surface acoustic wave
SC	switched capacitor
SCE	saturated calomel electrode
SEM	scanning electron microscopy
SHE	standard hydrogen electrode
SIMS	secondary-ion mass spectroscopy
SMT	surface mount technology
SS	stainless steel
SWV	square-wave voltammetry
TAB	tape automated bonding
TFS	thick-film sensor
TiW	titanium–tungsten alloy
UV	ultraviolet
VLSI	very large scale integration
WE	working electrode

Table A.4   List of subscripts

Symbol	Standard meaning
a	anodic
app	applied
c	cathodic
d	diffusion
dif	differential
f	(1) forward
	(2) faradaic
in	input
j	concerning species j
max	maximum
min	minimum
O	concerning the oxidized form O
out	output
p	peak
R	concerning the reduced form R
r	reverse
ref	reference

INDEX

304   *Index*

9 780367 402884